Lecture Notes in Mechanical Engineering

Lecture Notes in Mechanical Engineering (LNME) publishes the latest developments in Mechanical Engineering—quickly, informally and with high quality. Original research reported in proceedings and post-proceedings represents the core of LNME. Volumes published in LNME embrace all aspects, subfields and new challenges of mechanical engineering. Topics in the series include:

- Engineering Design
- Machinery and Machine Elements
- Mechanical Structures and Stress Analysis
- Automotive Engineering
- Engine Technology
- Aerospace Technology and Astronautics
- Nanotechnology and Microengineering
- Control, Robotics, Mechatronics
- MEMS
- Theoretical and Applied Mechanics
- Dynamical Systems, Control
- Fluid Mechanics
- Engineering Thermodynamics, Heat and Mass Transfer
- Manufacturing
- Precision Engineering, Instrumentation, Measurement
- Materials Engineering
- Tribology and Surface Technology

To submit a proposal or request further information, please contact the Springer Editor of your location:

China: Ms. Ella Zhang at ella.zhang@springer.com
India: Priya Vyas at priya.vyas@springer.com
Rest of Asia, Australia, New Zealand: Swati Meherishi at swati.meherishi@springer.com
All other countries: Dr. Leontina Di Cecco at Leontina.dicecco@springer.com

To submit a proposal for a monograph, please check our Springer Tracts in Mechanical Engineering at https://link.springer.com/bookseries/11693 or contact Leontina.dicecco@springer.com

Indexed by SCOPUS. All books published in the series are submitted for consideration in Web of Science.

More information about this series at https://link.springer.com/bookseries/11236

Giuseppe Conte · Olivier Sename
Editors

Proceedings of 10th International Conference on Mechatronics and Control Engineering

ICMCE 2021

Springer

Editors
Giuseppe Conte
Department of Information Engineering
Marche Polytechnic University
Ancona, Italy

Olivier Sename
GIPSA-lab
University of Grenoble Alpes
Grenoble, France

ISSN 2195-4356 ISSN 2195-4364 (electronic)
Lecture Notes in Mechanical Engineering
ISBN 978-981-19-1539-0 ISBN 978-981-19-1540-6 (eBook)
https://doi.org/10.1007/978-981-19-1540-6

This Springer imprint is published by the registered company Springer Nature Singapore Pte Ltd.
The registered company address is: 152 Beach Road, #21-01/04 Gateway East, Singapore 189721, Singapore

Preface

The 10th International Conference on Mechatronics and Control Engineering ICMCE-2021 was held as a virtual event on July 26–28, 2021. Researchers from all over the world attended the Conference from home and saw the live broadcast of the invited and contributed speeches.

Keynote lectures were given by Maria Pia Fanti (New green delivery and logistics strategies); Roberto Setola (Identification of Critical Nodes in Critical Infrastructures); Olivier Sename (Design and experimental validation of robust control approaches for mechatronic systems); Maciej Michałek (Modeling and Control of Articulated Vehicles: the Challenge and Beauty); Kheng Lim Goh (Towards an uncomplicated process for repairing damaged advanced fiber-placement composites: fundamentals, prototypes, sustainability). On behalf of the organizer of the Conference, I would like to thank the keynote speakers for having accepted the invitation. Their talks addressed problems in logistics, safety of infrastructures, mechatronics systems and manufacturing that have a large impact on the organization of our modern society and on the use of resources in industrial production.

The contributed papers covered a large area in mechatronics and control engineering, ranging from classical control theory and optimization to applications in electronic and mechanical engineering and in manufacturing and material science.

The success of the Conference is due to the high scientific level of all the contributions as well as to the enthusiasm of all the attendees. The member of the Technical Committee did a precious job in selecting the papers that were accepted for presentation and for inclusion in this book of proceedings among those that were initially submitted.

Besides the authors, the attendees and the TC members, I would like to thank the members of the International Advisory Committee Pierre Borne and Maria Pia Fanti, the Program Chairmen Olivier Sename and Roberto Setola, the Program Co-Chairmen Maciej Michałek and João Manuel R. S. Tavares and the Publicity Chairmen Ke-Lin Du and Sergei Alexandrov, whose work was fundamental for assuring the smooth running of the Conference.

Finally, I would like to thank the Conference Organizers for their efforts in assuring the success of ICMCE in the last 10 years and for giving me the opportunity to Chair this Conference.

Ancona, Italy
July 2021

Giuseppe Conte
Conference Chair

Contents

About the Editors

Giuseppe Conte received the Laurea with Honor in Mathematics in 1974 from the University of Genova. He has been Lecturer and Associated Professor at the University of Genova and, from 1990 to 2021, Professor of Automation at the Università Politecnica delle Marche, Italy. He has been Visiting Professor at the Ohio State University of Columbus, Ohio, in 1987 and at the Ecole Centrale de Nantes, France, in 1988. He has got a Fulbright Scholarship (1980, 1987) and a NATO Senior Fellowship (1987). He has been Chairman of the IFAC TC on Linear Control Systems (2011-2017); founders and Chairman of the Italian Chapter of the IEEE - Control Systems Society (1998-2015). He has been IPC Chairman and/or General Chairman and invited/keynote speaker in several international conferences. He has been Principal Investigator/Coordinator of international, EU and national funded research projects. He has been Associate Editors and Editor of Special Issues of journals in system and control theory. He has published about 400 papers in international journals and conference proceedings; he co-authored one book and edited four books of proceedings. He holds two Italian patents in the field of advanced sensory systems. His research interests are in algebraic/geometric methods in system and control theory and in robotics.

Olivier Sename received a Ph.D. degree from Ecole Centrale Nantes in 1994. He is now Professor at the Institut Polytechnique de Grenoble within GIPSA-lab. His main research interests include Linear Parameter Varying systems and automotive applications. He is the (co-)author of 3 books, 80 international journal papers, and more than 250 international conference papers. He was the General Chair of the IFAC Joint Conference SSSC-TDS-FDA 2013, of the 1st IFAC Workshop on Linear Parameter Varying Systems 2015 and he was the IPC Chair of the 2nd and 4th IFAC Workshop LPVS 2018 & 2022. He was Program Chair of the 10th International Conference on Mechatronics and Control Engineering in 2021 and 2022. He presented several plenary talks (IFAC SSSC 2019, IFAC LPVS 2021, ICMCE 2021, ICSTCC2015). He is Associated editor of the IEEE CSS Letters and of the European

Control Conferences. He is member of the IFAC TC 2.1 Linear Control Systems and 7.1 Automotive Control. He has led several industrial (Renault, Volvo Trucks, JTEKT, Delphi) and international (Mexico, Italy, Hungary, Spain) collaboration projects. He has supervised 34 Ph.D. students.

A Numerical Study on Phase Transformation and Hardness Evolution During Scanning Induction Hardening

M. Areitioaurtena, U. Segurajauregi, R. Hidalgo, I. Urresti, and E. Ukar

Abstract Induction hardening is a heat treatment that is frequently used to improve the mechanical properties of the surface of components typically subjected to contact loads, such as bearings or gears, extending component life. The simulation of the induction hardening process is very complex and time-consuming, which increases time-to-market in new industrial processes as the typical approach to define the induction hardening parameters is trial and error. In this work, we present a numerical study on the evolution of microstructural phases and hardness prediction for a complex industrial case, where the induction hardening process of a large-size pitch bearing is simulated. This kind of component is hardened by scanning, where there is relative movement between the inductor and workpiece and the inductor is followed by a quenching shower. Simulating scanning induction hardening is especially challenging as simultaneous heating and cooling occur, increasing the multiphysical interactions that must be simulated. The results show the evolution of the microstructure during induction heating and subsequent quenching accompanied by a prediction on the hardened case. A homogeneous hardened within the industrial span can be achieved in the workpiece after the combined static-scanning hardening is applied in the study.

Keywords Heat treatment · Large-size bearing · Process simulation · Scanning induction hardening · Coupled model · Multiphysics

---

M. Areitioaurtena (✉) · U. Segurajauregi · R. Hidalgo · I. Urresti
Ikerlan Technology Research Centre, Basque Research and Technology Alliance (BRTA), Paseo J.M. Arizmendiarrieta 2, 20500 Arrasate-Mondragon, Spain
e-mail: mareitioaurtena@ikerlan.es

E. Ukar
Department of Mechanical Engineering, University of the Basque Country, Alameda Urquijo S/N, 48013 Bilbao, Spain

G. Conte and O. Sename (eds.), *Proceedings of 10th International Conference on Mechatronics and Control Engineering*, Lecture Notes in Mechanical Engineering,
https://doi.org/10.1007/978-981-19-1540-6_1

1 Introduction

Induction hardening is a superficial heat treatment that is highly effective, integrable, quick, and non-contact and is usually applied to load-bearing and wear susceptible components. The required set-up for induction heating is very basic and requires the billet itself, a coil or inductor, and a source of current. The part is positioned near the inductor and heated by the eddy currents produced due to the electromagnetic phenomena [1]. The surface layer or other specific areas of the component are heated until the target areas are above austenization temperature A_3 and austenite is formed. Subsequent quenching is applied, usually by spraying water or aqueous polymers onto the hot surface or by immersion. The rapid cooling turns austenite into martensite, which is a microstructural phase that is much tougher and harder. The localized heating during induction hardening allows the core to remain ductile while a hard martensitic layer is reached on the surface. The hardened case depth and pattern can differ tremendously as they depend on the applied frequency current, which can range from 50 Hz to multiple megahertz, creating depths that can vary from less than one millimeter to the component's total volume in some cases.

The induction hardening process involves a variety of physics, which interact in a highly coupled way because the properties of the materials depend on several aspects, such as temperature and microstructure [2]. Due to its numerous and complex multiphysical interactions shown in Fig. 1, and the required vast amount of difficult-to-obtain data, the simulation of the induction hardening process has not been widely studied in academia, as these aspects increase computational time enormously, which leads to very little use of the current simulation tools in the industry. It is possible to test various process strategies and parameters (current intensity, frequency, power, scanning speed, and geometries) by simulating the induction hardening process,

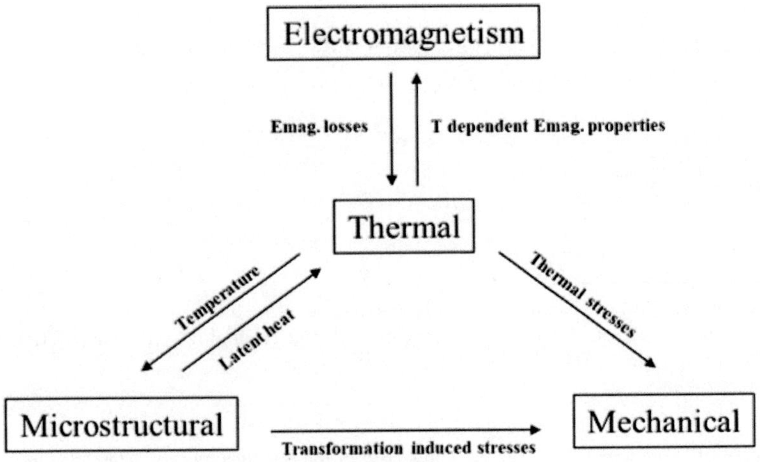

Fig. 1 Multiphysical interactions in induction hardening

in order to be able to predict final hardened case properties and also to determine possible process errors and failures, such as heating discontinuities, overheated areas, or cracking appearance, among others.

The electromagnetic–thermal coupling needs several back-and-forth iterations due to the sudden change in the electromagnetic properties of nonlinear ferromagnetic materials with temperature and is typically the coupling that requires the longest computational cycles, because electromagnetic simulation is extremely time-consuming.

Several published works have documented the simulation of the induction hardening process. A basic electromagnetic and thermal model is used in the work presented by Wrona et al. [3] to compute the evolution of temperature during the gear hardening by induction. Choi et al. [4] and Li et al. [5] carried out a similar analysis, where the prediction of the hardness profile in the cylindrical workpiece is included, achieving good agreement with experimental results. The research by Hömberg et al. [6] and Cajner et al. [7] show coupled models where microstructural transformations and hardness are computed and experimentally validated in spur-gears and discs and 42CrMo4 cylinders, respectively.

In this work a completely coupled electromagnetic-thermal-metallurgical-mechanical model is presented along an industrial case study, demonstrating that the developed model can be used to compute complex three-dimensional cases with adequate computational time, predicting the pattern and hardness values of the hardened case and its microstructure.

2 Modeling Scanning Induction Hardening

A fully coupled electromagnetic-thermal-microstructural-mechanical model is required for the simulation of the simultaneous scanning induction hardening process. In this work, the ANSYS commercial software has been used. The software is currently unable to compute microstructural transformations, but these can be incorporated into the model by implementing the constitutive equations into subroutines or using the software's own coding language, Ansys Parametric Design Language (APDL).

2.1 *Electromagnetic-Thermal Coupling*

A semi-analytical strategy has been used in this work to solve the electromagnetic coupling, as used in a previous work by the authors [8]. The usage of the semi-analytical approach brings shorter computational times than commercial dedicated softwares, especially for complex industrial cases where a three-dimensional computation is usually required. First, the electromagnetic solver resolves the diffusion Eq. (1) computed from Maxwell's equations. The computed magnetic field strength is introduced into an analytical model that, combined with a thermal FE model, is able to calculate the heat generation caused by the induced currents and the resulting temperature distribution, by resolving the heat Eq. (2).

$$\sigma\frac{\partial \boldsymbol{A}}{\partial t} - \frac{1}{\mu}\nabla^2\boldsymbol{A} = \boldsymbol{J}_S \tag{1}$$

$$\rho c_P\frac{\partial T}{\partial t} - \nabla(\kappa\nabla T) = \dot{q} \tag{2}$$

Further details of the semi-analytical approach and electromagnetic material properties can be found in Ref. [8].

2.2 Microstructure

During induction heating, γ austenite is generated when the temperature reaches the critical austenitization temperature range, where the initial microstructure is transformed between the austenite start and end temperatures (AC_1 and AC_3, respectively). For 42CrMo4, the critical temperatures that have been used in this work are 759 and 805 °C. Equation (3) can be used to describe the volumetric fraction of transformed austenite during heating [9].

$$f_\gamma = 1 - \exp\left(k\frac{AC_1 - T}{AC_1 - AC_3}\right) \tag{3}$$

The Johnson–Mehl–Avrami–Komogorov model, also known as JMAK model, describes the parent austenite decomposition into daughter phases when the transformation is considered diffusive, such as for the case of α ferrite, P pearlite, or α_b bainite, denoted as phase i in Eq. (4)

$$f_i(T,t) = 1 - exp\left[-b(T)t^{n(T)}\right] \tag{4}$$

where parameters $b(T)$ and $n(T)$ depend on the kinematics of each microstructural phase and are computed from the TTT diagram (these equations can be found in Ref. [10]).

Martensitic (α') transformation, which is the main microstructural phase generated during quenching, can be computed by Koistinen and Marburger Eq. (5).

$$f_{\alpha'}(T,t) = 1 - exp[-0.011(M_S - T)] \tag{5}$$

The critical temperatures for austenite and martensite transformations have been described by several analytical equations and depend on the chemical composition of the material.

Table 1 Chemical composition ranges of 42CrMo4 specified by standard EN 10083-3

	C	Mn	P	S	Si	Cr	Mo
Min	0.38	0.6	–	–	–	0.9	0.15
Max	0.45	0.9	0.025	0.035	0.4	1.2	0.3

$$A_1 = 723 - 10.7\ \text{Mn} - 16.9\ \text{Ni} + 29\ \text{Si} + 16.9\ \text{Cr} + 290\ \text{As} + 6.4\ \text{W} \quad (6)$$

$$A_3 = 912 - 203\sqrt{\text{C}} - 15.2\ \text{Ni} + 44.7\ \text{Si} + 104\ \text{V} + 31.5\ \text{Mo} + 13.1\ \text{W} + 30\ \text{Mn} - 11\ \text{Cr} - 20\ \text{Cu} + 700\ \text{P} + 400\ \text{Al} + 120\ \text{As} + 400\ \text{Ti} \quad (7)$$

$$M_S = 561 - 474\ \text{C} - 33\ \text{Mn} - 17\ \text{Ni} - 17\ \text{Cr} - 21\ \text{Mo} \quad (8)$$

The chemical composition of 42CrMo4 is specified by standard EN 10083-3. Typical values for 42CrMo4 are shown in Table 1.

A Time–Temperature–Transformation diagram (known as TTT diagram) is shown in Fig. 2, computed using the upper limits of the chemical composition of the steel specified by the standard.

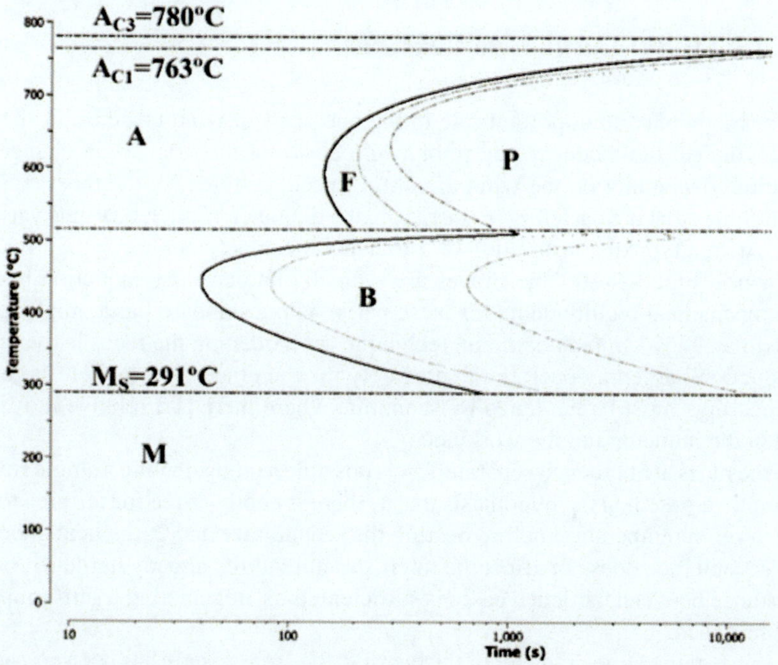

Fig. 2 Time–Temperature–Transformation diagram for 42CrMo4

2.3 Hardness

Vickers Hardness can be evaluated from the chemical composition of the material. Each microstructural phase has a specific hardness and a linear rule of mixture can be used to compute the hardness based on the fraction of microstructural phases and their hardness values.

$$HV_{\mathrm{M}} = 127 + 949\ \mathrm{C} + 27\ \mathrm{Si} + 11\ \mathrm{Mn} + 8\ \mathrm{Ni} + 16\ \mathrm{Cr} + 21 \ln V_r \quad (9)$$

$$\begin{aligned} HV_{\mathrm{B}} =& -323 + 185\ \mathrm{C} + 330\ \mathrm{Si} + 153\ \mathrm{Mn} + 65\ \mathrm{Ni} + 144\ \mathrm{Cr} + 191\ \mathrm{Mo} \\ &+ (89 + 53\ \mathrm{C} - 55\ \mathrm{Si} - 22\ \mathrm{Mn} - 10\ \mathrm{Ni} - 20\ \mathrm{Cr} - 33\ \mathrm{Mo}) \ln V_r \end{aligned} \quad (10)$$

$$\begin{aligned} HV_{\mathrm{F,P}} =& 42 + 223\ \mathrm{C} + 53\ \mathrm{Si} + 30\ \mathrm{Mn} + 12.6\ \mathrm{Ni} + 7\ \mathrm{Cr} + 19\ \mathrm{Mo} \\ &+ (10 - 19\ \mathrm{Si} + 4\ \mathrm{Ni} + 8\ \mathrm{Cr} + 130\ \mathrm{V}) \ln V_r \end{aligned} \quad (11)$$

where V_r denotes the cooling rate at 700 °C and chemical elements are introduced in weight percentage.

3 Case Study: Large-Size Bearings

Pitch bearings play an important role in the operation of wind turbines, as they are able to connect the blades to the rotor while allowing the orientation change that is required depending on the wind direction, speed, or force. In the most common wind turbines that are currently in operation, the diameter of pitch bearings varies up to 3 m, with a typical construction of a four-point contact double-row ball bearing, as shown in Fig. 3 (left). Their races are typically hardened by induction because of the mechanical requirements for these parts, with a standard hardening depth of approximately 5–7 mm. A common technique for hardening the races is to conduct the induction hardening operations separately for each race. Due to their large size, pitch bearings must be hardened by scanning, where there is a relative movement between the inductor and the workpiece.

As the races are hardened separately, it is possible to consider that a single row can be computed based on the hypothesis that 1: there is enough time for the area to cool down to room temperature before heating the second race and 2: the heat generated in the second race does not affect the microstructure of the already hardened race as the distance between hardened cases is sufficient, thus, not creating a self-tempering of the first race.

Because of the large diameter of the bearing, the area of study has been considered straight for the sake of simplicity. For the initial static electromagnetic simulation, a double-turn inductor is used, along a flux concentrator that allows more efficient

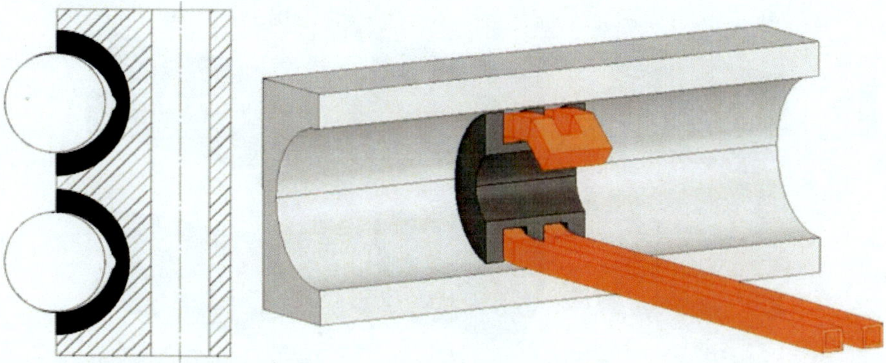

Fig. 3 Cross-section of a double row ball bearing (left) and electromagnetic model including inductor and field concentrator (right)

usage of the machine power—thus, improving power consumption and opening the possibility to use less powerful machines. The air must also be included in the electromagnetic FE simulation, giving a total of 100 k elements. For the subsequent coupled thermal-metallurgical-mechanical simulation, only the workpiece is modeled using 390 k elements, where the surface and subsurface areas have a very fine mesh in order to use the developed semi-analytical electromagnetic-thermal approach. Induction parameters are 20 kHz and 2100 A (RMS value) and the scanning speed is set to 3.5 mm/s. During the scanning process, the inductor is followed by a shower that projects an aqueous polymer that cools the hot surface fast enough to generate the martensitic layer. For the developed model, the shower is modeled as a "cooling box," where a high convection coefficient (15,000 W/m^{2}K) is placed in selected moving areas. The cooling box is placed with a separation of 20 mm from the inductor with a width of 20 mm. The total process time was set to 30 s and the typical computational time for the coupled simulation was 3 h using 8 cores on a server equipped with Intel® Xeon® Gold 6242 CPU dual-processor (2.8 GHz) and 512 Gb RAM.

Figure 4 shows the evolution of the austenitized region during the scanning process and the martensitic layer after quenching. In the figure, it can be observed that the austenite is fully transformed into martensite, obtaining a hardened layer of approximately 5–8 mm depth, the maximum hardness being 860HV.

The results show that 12 s are required for the system to reach homogeneous austenitization temperatures through the cross-section. Thus, an approximate length of 42 mm, which corresponds to the approximate width of the inductor, will remain unhardened using the proposed hardening strategy, which is usually called soft spot or soft zone. In order to reduce or totally avoid soft zones, one can find several patented techniques using one or more inductors.

One of the main results that are critical for service life is the residual stress pattern obtained after induction hardening. Further development of the presented model will include the prediction of residual stresses after the heat treatment, as it has been

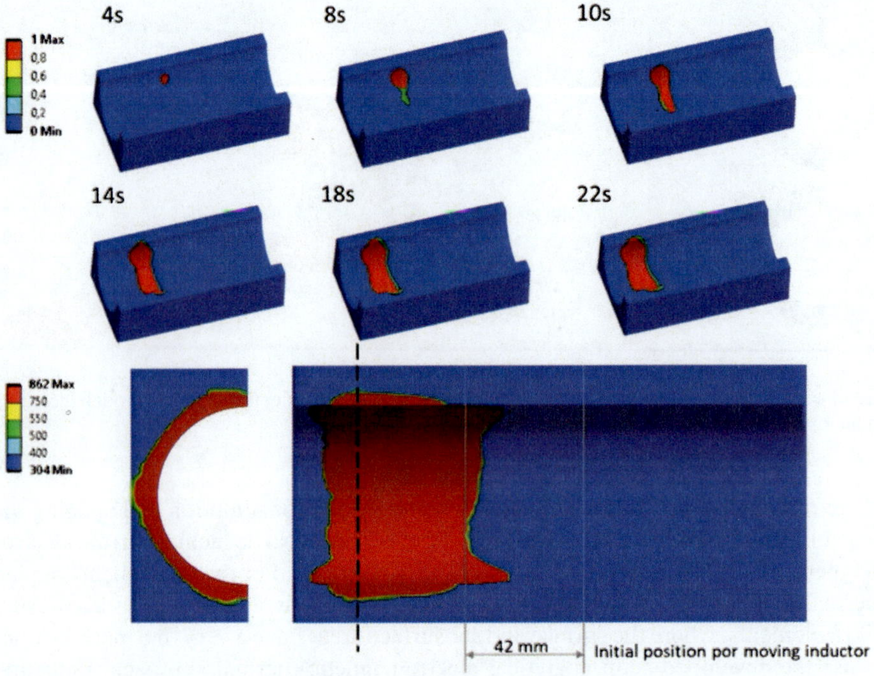

Fig. 4 Austenite evolution during scanning induction hardening (above) and hardness layer in workpiece and cross-section after 30 s (below)

demonstrated that compressive residual stresses on surface and subsurface areas slow down crack propagation, extending component life.

4 Conclusions

A coupled electromagnetic-thermal-metallurgical-mechanical model was presented in this paper for the numerical simulation of the scanning induction hardening process in complex industrial cases, which brings the possibility of carrying out extensive numerical analyzes in the industrial context, possibly reducing time-to-market and very high costs associated with the process definition developments.

The developed model was applied to an industrial-based case study, where the scanning induction hardening of large-size pitch bearings was simulated, obtaining expected results for microstructure and hardened depth and case pattern. In this field, it is of great importance to achieve an efficient and favorable process as structural integrity is highly affected by the results of the induction hardening process and can be critical for service life.

Acknowledgments Ikerlan's research has been supported by CDTI, depending by Ministerio de Ciencia e Innovación, through the "AYUDAS CERVERA PARA CENTROS TECNOLÓGICOS 2019" program, project MIRAGED with expedient number CER-20190001. This research center is certificated as CENTRO DE EXCELENCIA CERVERA.

References

1. Rudnev V, Loveless D, Cook RL (2017) Handbok of induction heating
2. Şimşir C (2014) Modeling and simulation of steel heat treatment—prediction of microstructure. Distortion, Residual Stress, Crack ASM Handb 4B:409–466
3. Wrona E, Nacke B, Resetov D (2003) 3D-modelling of the transient eating process for induction surface hardening MEP-international scientific colloquium modelling for electromagnetic processing, pp 119–123
4. Choi JK, Park KS, Lee SS (2018) Prediction of high-frequency induction hardening depth of an AISI 1045 specimen by finite element analysis and experiments. Int J Precis Eng Manuf 19(12):1821–1827
5. Li H, He L, Gai K, Jiang R, Zhang C, Li M (2015) Numerical simulation and experimental investigation on the induction hardening of a ball screw. Mater Des 87:863–876
6. Hömberg D, Liu Q, Montalvo-Urquizo, Nadolski D, Petzold T, Schmidt A, Schulz A (2016) Simulation of multi-frequency induction hardening including phase transitions and mechanical effects Mater Des 121:86–100
7. Cajner D, Smoljan B, Landek D (2004) Computer simulation of induction hardening. J Mater Process Technol 157–158:55–60
8. Areitioaurtena M, Segurajauregi U, Urresti I, Fisk M, Ukar E (2020) Predicting the induction hardened case in 42CrMo4 cylinders. Procedia CIRP 87:545–550
9. Xia J, Jin H (2018) Numerical modelling of coupling thermal metallurgical transformation phenomena of structural steel in the welding process. Adv Eng Softw 115:66–74
10. Carlone P, Palazzo GS (2011) Development and validation of a thermomechanical finite element model of the steel quenching process including solid-solid phase changes. Int Appl Mech 46(8):995–971

Novel Methodology for Induction Hardening Simulation of a Large Pitch Bearing

R. Hidalgo, M. Areitioaurtena, I. Hinojosa, U. Segurajauregi, and I. Urresti

Abstract Induction hardening is a complex process as multiphysics interactions occur. Due to that, it is also difficult to simulate because an accurate model which considers all the variables is computationally expensive. Thus, the definition of the parameters required for the process is based on the trial-and-error technique, which slows down and gets expensive the components. In this work, a new methodology to simulate the induction hardening process is presented and it is employed to design the induction hardening process required to obtain an adequate hardened layer in a 7 m diameter pitch bearing designed according to the expected requirements for the future wind turbines. The results obtained showed that a scanning inductor with an intensity of 12,000 A, a frequency of 2 kHz and a scanning speed of 6 mm/s meet the requirements. These results were obtained in a reasonable time. Therefore, this work may entail an important milestone towards the implementation of induction hardening simulation in the industry.

Keywords Induction Hardening · Large-size bearing · Manufacturing process simulation

1 Introduction

Large bearings are employed in wind turbines and other energy industry applications where they are subjected to harsh working conditions. Raceway surface is treated to obtain a hardened layer while the core material maintains good ductility. This layer is responsible for carrying the high dynamic loads the bearing undergoes. To obtain this layer, the workpiece is typically subjected to induction hardening.

R. Hidalgo (✉) · M. Areitioaurtena · I. Hinojosa · U. Segurajauregi · I. Urresti
Ikerlan Technology Research Centre, Basque Research and Technology Alliance (BRTA), Arrasate-Mondragon, Spain
e-mail: rhidalgo@ikerlan.es

M. Areitioaurtena
e-mail: mareitioaurtena@ikerlan.es

G. Conte and O. Sename (eds.), *Proceedings of 10th International Conference on Mechatronics and Control Engineering*, Lecture Notes in Mechanical Engineering,
https://doi.org/10.1007/978-981-19-1540-6_2

Induction hardening is a surface heat treating method in which a metal part is heated by induction and then cooled in a quenchant, such as water, polymer, oil or high-pressure gas. During induction hardening, the quenched metal undergoes a phase transformation of austenite to martensite and the hardness of the quenched part increases. The induction heating process uses the principle of electromagnetic induction to heat the surface layer of a workpiece. By placing a conductive material into a strong alternating magnetic field, an electrical current can be made to flow in the material, thereby creating heat due to the Joule effect in the material. An important material property of the ferromagnetic material is the Curie temperature or Curie point, which is the temperature where a material's permanent magnetism changes to induced magnetism. Thus, induction hardening is widely used to selectively harden the surface area of a part without affecting the interior mechanical performance of the component [1].

It is a complex heat treatment because many technological parameters, such as the size and form of the induction coil, current frequency, voltage, current, moving rate of the induction coil, the radial distance between the workpiece and induction coil, and so on, can affect the resulted hardened layer, and it is difficult to ascertain and select these parameters in induction hardening [2], as many specific commercial software require a high computational time. Thus, the procedure for designing an induction heating process in the industry relies on the experts' experience, trial-and-error and company know-how, since the numerical simulation of the induction heating process is not widely spread in the industry because of its complexity and hard-to-obtain material properties [3].

Regarding literature, most of the authors developed analytical models applied to simple 2D or axisymmetrical geometries [4, 5]. Others simplify the problem by developing a one-way workflow in ANSYS omitting the thermal dependence of electromagnetic properties [6] or using linear materials such as aluminium [4]. The use of ferromagnetic materials as well as 3D geometries increase computational time. Thus, 3D computation of induction heating has not been widely studied.

To overcome this issue, in this paper, a new methodology to simulate induction hardening of nonlinear materials with reduced computational time is presented. To demonstrate its applicability, it was used to find the process parameters of the induction hardening process of a large pitch bearing.

2 Case Study

The case study used in this paper is based on an innovative pitch bearing concept that will be installed in a 20 MW offshore wind turbine from the year 2030–2050 onwards (hub height of 160, rotor diameter of 276 m and the pitch bearing diameter of 7 m) (Fig. 1a). The turbine will be installed on a wind farm with a size of 2 GW with 100 turbines. The pitch bearing design is based on a 3 row roller bearing (3RRB) and was developed within the INNTERESING project [7].

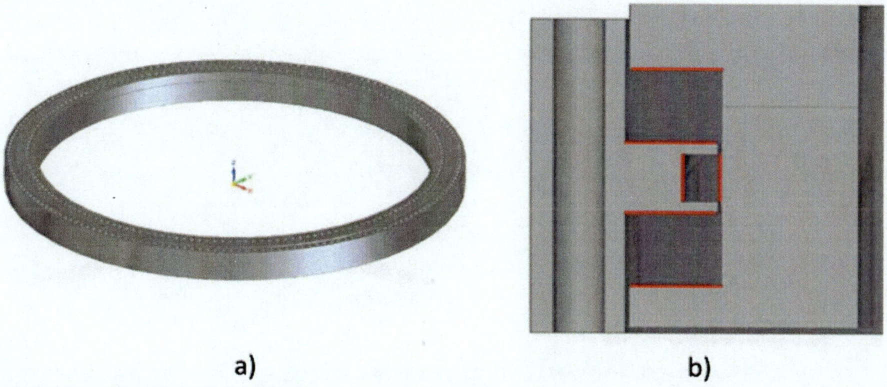

Fig. 1 **a** Design of the 3RRB for 20 MW wind turbine and **b** Detailed view of the zones to be hardened

In the case study of the present paper, the zones to harden are highlighted in Fig. 1. Based on the authors' knowledge, the horizontal layer of the inner ring is more critical against fatigue failure, for this reason, the analysis will centre on this zone. Additionally, to simplify the procedure, two simplifications were assumed:

- Only a section of the ring was considered for the calculation because, as the layer will be constant for all the ring, the results of a small section will reflect the whole model.
- The section is considered straight. As the diameter of the ring is 7 m, the curvature of the section can be considered negligible. This simplifies the movement equations, so linear movement can be used instead of circular.

Figure 2 shows the target layer to harden in the section to analyse. This coincides with the railway of the rolling elements. Additionally, the zone where the horizontal and vertical faces meet must be hardened because it promotes stresses concentration. The minimum depth of this layer must be 10% of the diameter of the rolling elements: 12.5 mm.

Although new materials with higher properties can be developed in the upcoming years, for this approach, 42CrMo4 steel was used as it is the material that is currently used to manufacture this type of bearings.

3 Methodology

The induction hardening process implies interaction between electromagnetic and thermal phenomena because the magnetic properties of the materials vary with the temperature. Additionally, the changes in the temperature along the process produce microstructural transformations, stresses, distortions and variation of the hardness. For this reason, the simulation of this process is complex and computationally

Fig. 2 Target layer to harden

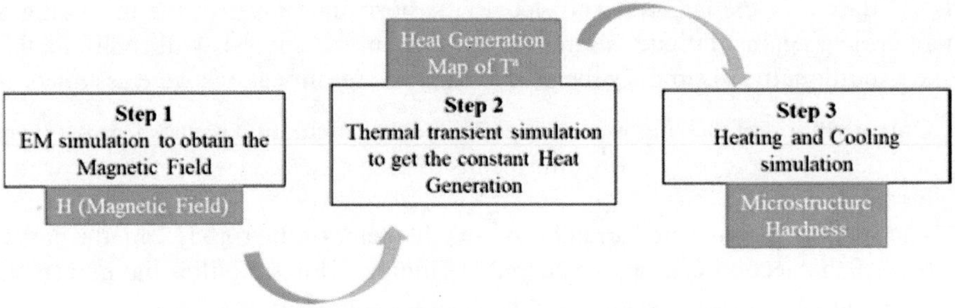

Fig. 3 Developed methodology to simulate the induction hardening process

expensive. The proposed methodology consists of three steps that are shown on Fig. 3:

- Electromagnetic simulation. One electromagnetic simulation is needed to obtain a map of the magnetic field strength that will be the input for the next step.
- Thermal simulation. The map with the magnetic field strength is introduced in a thermal transient simulation in which an ANSYS-APDL code is implemented to calculate, by means of a semi-analytical approach, the stable heat generation and the map of temperatures at the end of the simulation will be the input for the next step. The scanning speed of the inductor is considered.
- Heating and cooling simulation. The last step combines heating and cooling in movement and, as a result, the microstructure and hardness after the process are obtained.

3.1 Electromagnetic Simulation

In the presented approach, only one simulation regarding electromagnetic physic is required. The geometry of the bearing, the inductor and the surrounding media are introduced in the commercial finite element software ANSYS Maxwell, Release 2020R1. From this simulation, the magnetic field strength in the workpiece at room temperature is obtained.

3.2 Thermal Simulation

In this work, the semi-analytical induction heating model presented in [8, 9] is used to compute the induction hardening process. It was adapted to be applied to 3D geometries and to include the movement of the inductor. The use of the semi-analytical model allows the computational time to be much faster compared to a fully coupled model using commercial software.

As the hardened layer in the pitch bearing is constant along the raceway, it is assumed that, after a time at a constant speed, the heat generated in the workpiece stabilized and become constant. This constant heat and the temperature map are the inputs for the next step.

3.3 Heating and Cooling Simulation

In this step, the austenite formation during heating is calculated by setting the corresponding AC1 y AC3 as 760 C y 805 °C, respectively, for 42CrMo4. Then, during cooling, the austenite formed during heating transforms into martensite below the Ms temperature. Additionally, the hardness of the material after cooling is evaluated based on the fraction and hardness of each phase in each point. All these results can be calculated by developing user subroutines and including them in ANSYS.

4 Application to the Case Study

The above-explained methodology was applied to find the process parameters require in the future to get an adequate hardened layer for the presented case study. During the induction hardening of pitch bearings, the coil and the water spray scan around the bearing. For this reason, apart from the power and frequency of the system, the speed of the inductor is also an important parameter in the process as it determines the depth and quality of the hardened layer.

First of all, the geometry of the induction system must be designed. To this end, different geometries were simulated in ANSYS Maxwell. Due to the limitations of the software, instead of power, intensity values were used. The intensity was 12000A and the frequency 2 kHz. The obtained magnetic field strength for each case was introduced in the thermal simulation to obtain the resulting temperature map after a fixed time of 20 s. As a first approximation, it was considered that the zones whose temperatures were above 805 °C would completely transform into martensite during cooling. For that reason, the pattern of the temperature map above this value must be as similar as possible to the pattern shown in Fig. 2.

Once the geometry of the inductor is selected, two intensities and three velocities are tested in order to find the most optimized process. That is to say, the fastest speed and the lowest intensity at which the required hardened layer is obtained. Each simulation consisted of keeping the inductor fixed for 7 s and moving during 25 s at each velocity. The point to consider the hardened layer obtained was the one at which the heat generation becomes stable.

5 Results

The first step was to design and test different inductor system configurations (Table 1). The inductor system consists of a copper inductor and an additional flux concentrator. The changes were centred in the latter with the objective of getting a uniform magnetic field strength on the surface. This uniform magnetic field will produce higher temperatures in the surfaces. Over time, the heat generation will go deeper into the workpiece due to the thermal conduction and varying magnetic permeability. The resulting temperature map after 20 s in static and calculated with the semi-analytical model was evaluated. Comparing these maps with the expected hardened layer required, geometry C was selected because the maximum temperatures were distributed along a higher part of the surface and the interior.

Once the geometry of the induction system was selected, the next step is finding the optimal parameters of the process. To this end, two intensities were set; 12,000 and 10,000 A and three speeds of the inductor; 4, 6 and 8 mm/s. The temperature map in a section plane is shown for each case in Table 2. As can be seen, the most suitable induction parameters to use are 12,000 A and 6 mm/s as the required layer is obtained and the speed of the process can be increased with respect to the velocity of 4 mm/s, which also produced an adequate temperature pattern.

Table 1 Inductor system proposals and resulting temperature map after 20 s

A	B	C
	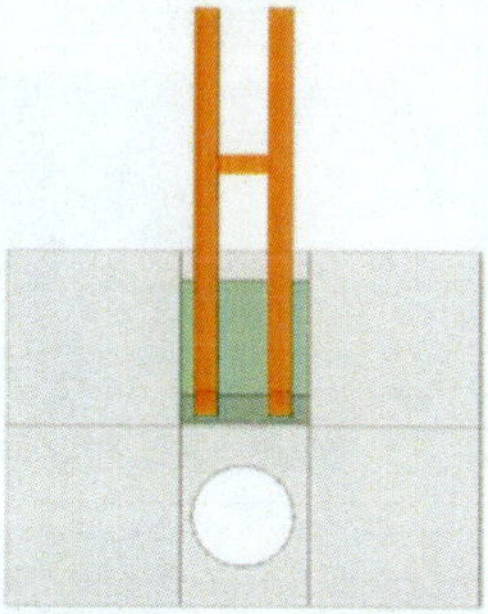	
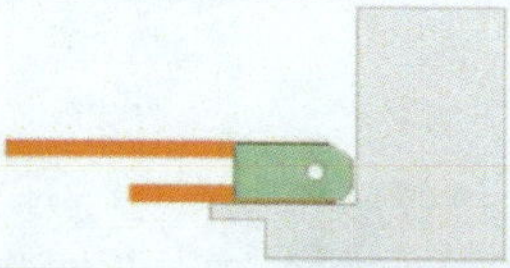	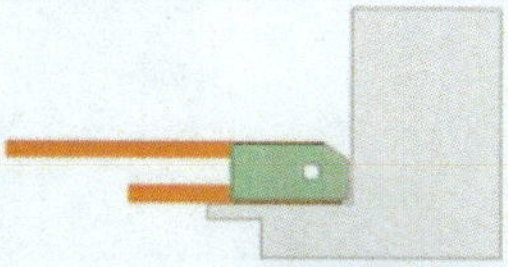	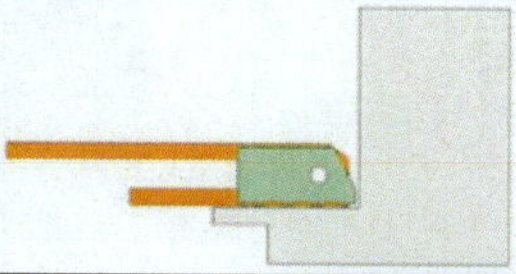
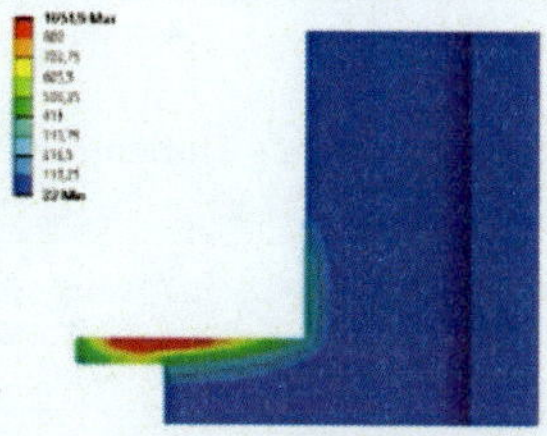	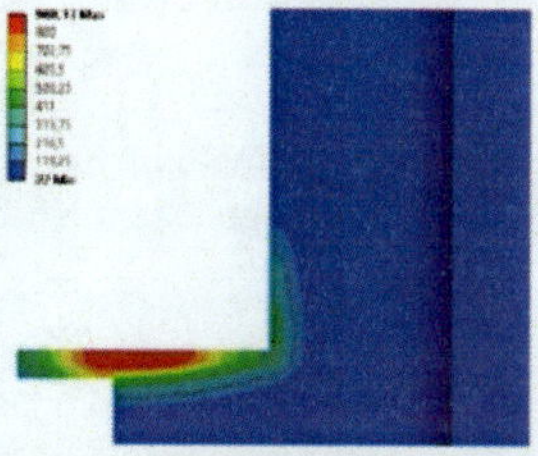	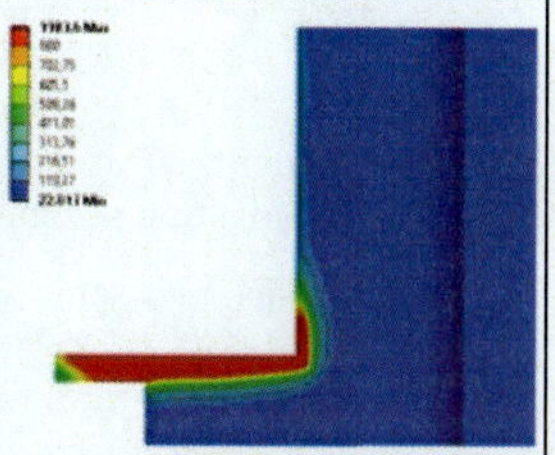
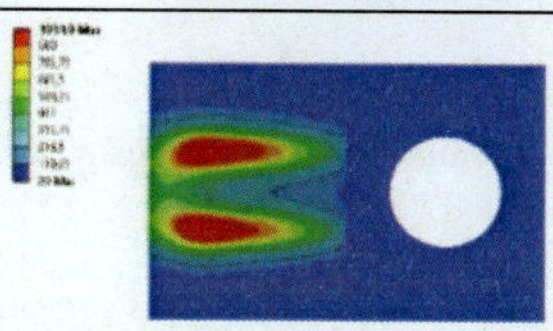	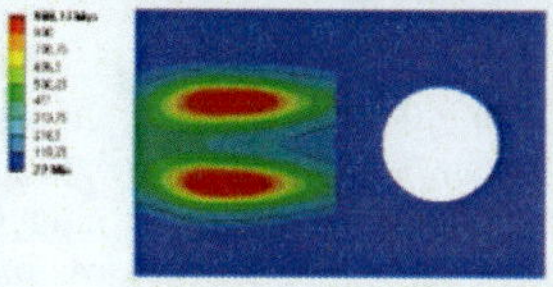	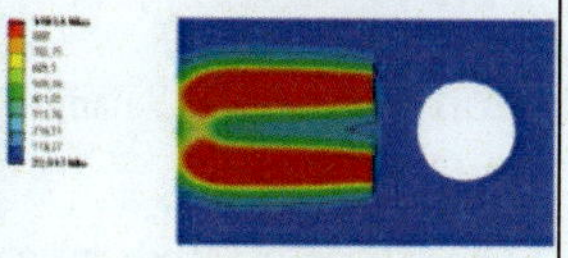

6 Conclusions

A new methodology for simulating the induction hardening process was presented and applied to find the process parameters of a 7 m of diameter pitch bearing. Different induction systems geometries were tested and the one that provides the most superficial area at a higher temperature was selected. Then, it was found that an intensity of 12,000 A, a frequency of 2 kHz and a scanning speed of 6 mm/s provided an adequate hardened layer.

This work entails an important milestone towards the implementation of induction hardening simulation in the industry because the times required to simulate the complete process are quite reasonable.

Table 2 Resulting temperature map in a section plane depending on the current intensity and speed of inductor

	V=4mm/s	V=6mm/s	V=8mm/s
I=10000 A	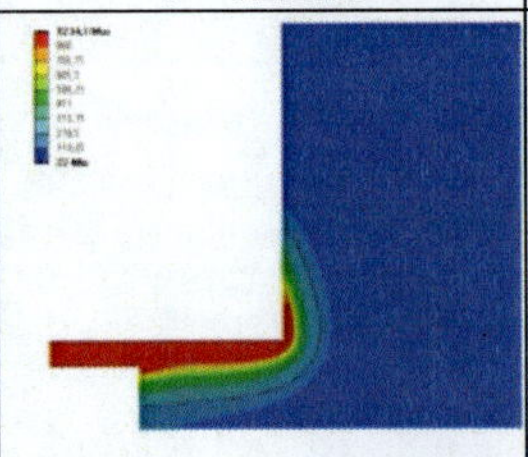	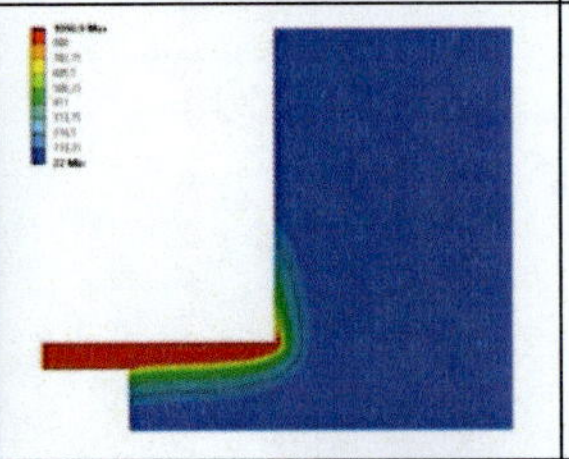	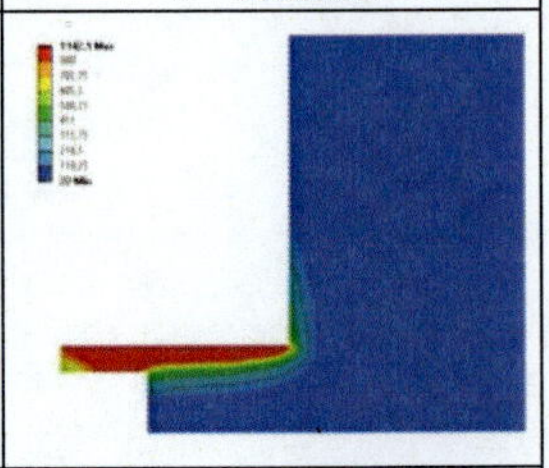
I=12000 A	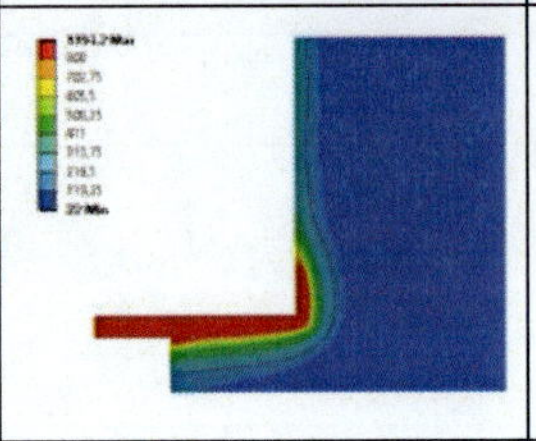	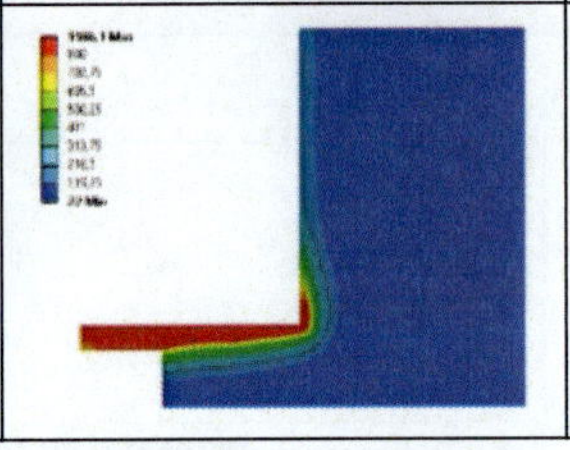	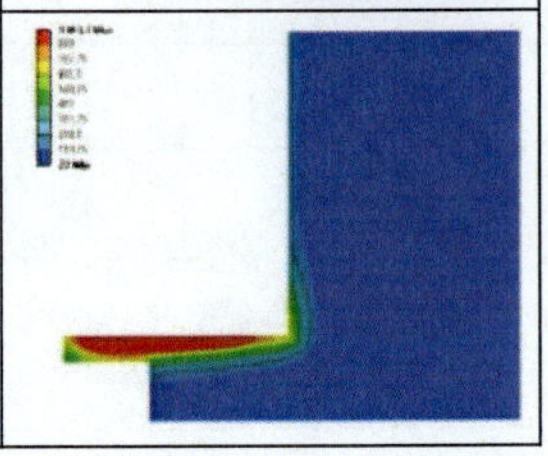

Acknowledgements This project has received funding from European Union's Horizon 2020 research and innovation programme under grant agreement No 768772.

References

1. Li H, He L, Gai K, Jiang R, Zhang C, Li M (2015) Numerical simulation and experimental investigation on the induction hardening of a ball screw. Mater Des 87:863–876
2. Rapoport E, Pleshivtseva Y (2006) Optimal control of induction heating processes
3. Simsir C (2014) Modeling and simulation of steel heat treatment—prediction of microstructure, distortion, residual stresses, and cracking. In: ASM handbook, steel heat treating technologies, vol 4B
4. Kennedy M, Akhtar S, Bakken J, Aune R (2011) Analytical and experimental validation of electromagnetic simulations using COMSOL®, re inductance, induction heating and magnetic fields. In: COMSOL users conference, Stuttgart Germany
5. Bay F, Labbé V, Favennec Y, Chenot JL (2003) A numerical model for induction heating processes coupling electromagnetism and thermomechanics. Int J Numer Methods Eng 58:839–867
6. Zhang X, Chen C, Liu Y (2015) Numerical analysis and experimental research of triangle induction heating of the rolled plate. Proc Inst Mech Eng Part C J Mech Eng Sci 231
7. https://www.innterestingproject.eu/
8. Areitioaurtena M, Segurajauregi U, Urresti I, Fisk M, Ukar E (2020) Predicting the induction hardened case in 42CrMo4 cylinders. Procedia CIRP 87:545–550
9. Areitioaurtena M, Segurajauregi U, Akujärvi V, Fisk M, Urresti I, Ukar E (2021) A semi-analytical coupled simulation approach for induction heating. Adv Model Simul Eng Sci 8(1). https://doi.org/10.1186/s40323-021-00199-0

Review and Methodology on Vision-based Sensing Approach in Metal Additive Manufacturing Process

Natago Guilé Mbodj and Peter Plapper

Abstract Additive Manufacturing (AM) of metallic objects is the process of adding layer-upon-layer of material to produce a hard-compact physical object. During the deposition, many process parameters, such as thermal history, power, material feeding rate, microstructure, etc., are involved and need to be regulated. The soar of the 3D printing process in various sectors has recently added a set of challenges, creating a massive demand for an automated process monitor to ensure a good deposition. In this paper, a literature review of vision-based techniques in additive manufacturing process is presented. A molten pool profile extraction methodology is performed. The image processing results can be used to create a visual monitoring technique and inspire future research in the use of vision sensing in 3D printing process.

Keywords 3D printing · Vision sensing · Weld pool · Profile extraction

1 Introduction

In recent years, Additive Manufacturing (AM) has been developed in terms of design and functions. 3D printing of objects has increased rapidly and become more open to the public. Additive Manufacturing is simply the appellation for rapid prototyping, also called 3D printing [1]. The principle of AM process is to add material layer by layer. A layer symbolized a cross-section of the material obtained from a Computer-Aided Design (CAD) model. Earlier, AM was used to visualize products as they were being developed. Today, some end-use products for automobiles, aircraft, medical implants, dental restoration, and even fashion products are manufactured using AM techniques [2]. The combination of AM with robotics permits the manufacture of

N. G. Mbodj (✉) · P. Plapper
University of Luxembourg, 6, rue Coudenhove-Kalergi, L-1359 Esch-sur-Alzette, Luxembourg
e-mail: natago.mbodj@uni.lu

P. Plapper
e-mail: peter.plapper@uni.lu

G. Conte and O. Sename (eds.), *Proceedings of 10th International Conference on Mechatronics and Control Engineering*, Lecture Notes in Mechanical Engineering,
https://doi.org/10.1007/978-981-19-1540-6_3

larger scale objects. A 6 DOF robot manipulator offers more flexibility and workspace compared to 3D printers. Robotics additive manufacturing process has the following advantages [3, 4]: (1) Product manufacturing time is significantly reduced; (2) Since parts are not cut from a larger block, material waste is nearly reduced to zero; (3) Complex large parts are now able to be manufactured; (4) Finally, the automatization of the process comes up with less supervision in the working environment.

AM integration in the industrial sector aims to create high-quality complex parts in a large number. However, a new set of problems arises. The geometry, substrate temperature, cooling rate, power, speed, and many other factors are identified to influence the deposition quality [5]. The "teach and playback" robots do not meet the requirements to ensure a stable deposition. The process is usually open-looped with process parameters fixed in advance. Thus, any unexpected change in the deposition environment results in a weld seam quality change and affects the final part accuracy. Therefore, the "teach and playback" method cannot automatically regulate the deposition process, leading to poor weld seam quality [6]. Indeed, this method is still widely used in industrial production. Today's objective is to replace "teach and playback" robots to automatically monitor the surrounding environment and correct real-time disturbances in the deposition process. Therefore, an autonomous system is essential to fill the existing shortages with this open-loop system. 3D printing is MIMO (Multiple-Input Multiple-Output), time-varying, nolinear, with multiple parameters affecting the deposition such as the velocity, wire feeding velocity, shielding gas, power or current, welding direction [7, 8]. A wrong combination or variation of these parameters leads to disturbances, thus induces defects in the produced part [9, 10], thus the motivation to have a sensor-based system [11]. The selected sensor shows a preview of the future deposited part. For instance, a pyrometer can only measure the temperature, whereas a camera provides a larger spectrum of deposition features. In fact, it is crucial to choose a sensor displaying the dynamics of the process and the weld pool characteristics. We believe that a vision-based system offers the requirements to extract weld pool information that can be used later for a real-time feedback correction system of the parameters to achieve better product accuracy.

The paper provides some key advantages of vision-based systems in AM and aims to inspire future research in vision-based monitoring in the AM industry. In Sect. 2, a review of vision-based is presented. A profile extraction is developed, and a methodology is proposed in Sect. 3; Sect. 4 concludes by summarizing this paper's findings.

2 Vision-based Sensing Techniques

In the literature, many works related to visual sensing to improve AM process have been reported. We begin this section by citing some of them before investigating the vision-based systems. The works treated in the literature include sensing features such as layer height, bead geometry, thermal history, etc.

In [9], Mazumder et al. investigated and analyzed independent parameters on the fabricated parts. Logic gates were used to combine the photodetector signals before a feedback system received them to control the deposit height. It is found that increasing the number of photodetectors in the feedback controller eliminated the directional dependence of the height controller on the material deposition, increased the material density slightly, and reduced the surfaces' roughness by an average of 14–20%. However, the use of multiple sensors does not have major effects on the material properties of the fabricated parts that need to be considered to improve the deposited parts' accuracy.

In [12], Bi et al. studied the influence of using different process control strategies on the dimensional accuracy, microstructure, and hardness of the final part. An infrared (IR)-temperature generated signal from the melt pool and is connected to a PID controller that regulates the laser power to control the melt pool temperature. Thin walls were deposited and compared under different processing conditions. It is found that with a constant set-value used in the process control, the quality of the deposited parts improves slightly. However, some discontinuities were still visible at the part's edges due to low cooling time. Therefore, preset values must work into a certain threshold to compensate for melt pool variation.

In [13], Toyserkani et al. developed a feedback control system to enhance the deposition quality. A CCD camera provides pool images. A pattern recognition algorithm is used to extract the height (z) and liquid/solid interface dimensions and angle in real-time. The camera is connected to a knowledge-based PID controller that adjusts the laser pulse energy. The deposition performed using this approach showed an improvement in the accuracy and overcame disturbances around the operating point. However, this approach is limited to simple straight deposition, not to curvature or complex deposition processes.

In [14], Cheng et al. presented a closed-loop control model that measures the final part's defects. A CCD camera with a Gaussian function is used as a shape descriptor and takes the surface layer's physical characteristics. The measurements are then fed to different process dynamics models designed in SIMULINK to compute the next layer's deposition flow rate compensation. The results show some accuracy in the reconstruction and detection of patterned surface shapes. However, the proposed approach reconstructs the 3D surface by using approximation and ignores other variations occurring in the deposition process.

Heralić et al. [15] developed a monitoring system to control the deposition process in real-time, mainly the bead height and width. Two cameras and a projected laser line are combined with an integrated PI controller and a feed-forward compensator for the bead width and the bead height. The control signals were the wire feed rate and laser power. Single-input, single-output (SISO) models represented the relationship between the bead geometry and the control signals. Single bead walls were deposited, and the results show good deposition stability using this approach. However, Heralić method's is not fully automated as the start and stop sequences are manually performed. Also, automatic feedback control considering other environmental disturbances in the deposition process can enhance the approach's robustness.

In [16], Doubenskaia et al. investigated the laser impact zone temperature during the deposition process. A bicolor pyrometer is used to measure the temperature and a CCD camera integrated with a PHENIX PM-100 machine to monitor the thermal radiation. Experiments were conducted under variable operational parameters (hatch distance, printing velocity, layer), and the result showed that the recorded signals are sensitive to the variation of the process parameters. However, as AM is complex, a pyrometer cannot interpret laser impact Zone as other environmental factors influence the pool surrounding.

Xiong et al. [17] developed a real-time measurement for the bead geometry (width and height). Two cameras to capture the images, a Gaussian filter to remove noises, a Sobel operator to detect the edge of the bead, and Hough transform algorithms for curve fitting composed the system. The goal was to monitor the nozzle to the top surface distance to eliminate the height errors during the deposition process. The validation tests showed the reliability of the vision-based online measurement. However, the main process parameters' correlation needs to be considered to develop a robust feedback control system using this approach.

In [18], Ocylok et al. investigated the process parameters' influence on the melt pool geometry. A camera is used to measure the melt pool's length, area, and diameter at different process parameters. The values are compared with the cross-sections and the laser spot diameter. An analysis software performs the calculation for each of the images. The experimental results of Ocylok et al.'s work show that the camera provided accurate information of the weld seam and allowed the detection of variations of process parameters influencing the melt pool size. However, it is difficult to observe other process parameters' real influence except for the power with this approach.

In [19], Grobert et al. developed a methodology to detect discontinuities in the deposition. A high-resolution digital single-lens reflex (DSLR) camera is used to collect multiple images at each build layer. Such linear Support Vector Machines (SVM) technique used a binary classification technique that detects flaws in the images. Two build structures (flaw and normal) were used to classify the process conditions. A computed tomography (CT) scan is used to label each layer. The images generated are permitted to identify fusion, porosity, and inclusions with automated analysis tools or manual inspection. Cross-validation experiments reveal that the camera-based system gives an accuracy greater than 80%. This approach can be more robust if it can detect near edges of parts and improve image resolution or with additional sensors to manufacture complex geometries.

Clijsters et al. [20] studied online monitoring and estimation of the quality of the deposition. The system comprises a high-speed near-infrared (NIR) thermal CMOS camera and a photodiode connected to a field-programmable gate array (FPGA) to transfer the images at high sample rates of at least 10 kHz to the control unit. The measured melt pool data are transferred on a 2D map grid for analysis and interpretation using a mapping algorithm. The experiments show that this model's images display the melt pool variation, thus defects in real time. However, the detection resolution, speed, and other deposition parameters must be studied for a robust model.

In [21], Xiong et al. developed a vision-based real-time sensing that adjusts the wire feed speed in the next layer. A camera with composite filters is mounted to a robot connected to an external control unit. The images are processed with a Gaussian filter that removes the noise, performs edge detection, calculates the optimal threshold, and a Hough transformation algorithm for edge fitting. A PID controller adjusts the wire feed rate. Deposition of thin-walled allowed to validate the approach's effectiveness. However, the wire feed rate only, as a controlling signal, is not enough to guarantee deposition stability because the other process parameters, for instance, the power, has a larger influence in the fabrication process.

Summarizing, the selection of a sensor system in 3D printing process has to overcome some of the deposition challenges by offering high-quality information to build a reliable monitoring process.

3 Case Study of a Molten Pool Profile Extraction

Real-time sensing of additive manufacturing processes remains a challenging topic in modern manufacturing technology. The first step is to place ourselves in a welder's position to imitate and extract weld pool size (i.e., sensing of the printing process). Then, identify or develop a sensing device similar to human sensing systems. The sensor needs to detect some of the deposition process conditions and adapt to environmental variations. The sensing device should acquire the seam dynamics features and gives two-dimensional or three-dimensional information of the weld pool surface. Therefore, machine vision and extraction of weld seam geometric information are techniques that can be employed. A camera is convenient to capture the melt pool images of the process. CCD (Charge Coupled Device) is suitable for quality control of the 3D printing process [22], as it provides sufficient information of the weld seam such as the position of the wire, the shape of the pool as well as the shape of the solidified deposition, etc. Compared with other sensing devices, a camera is contactless and the signal detection does not affect the printing process.

After the images are taken, an image processing algorithm is required [23]. The image processing algorithms need to adapt to some changing conditions. Many existing and efficient algorithms exist, such as curve fitting, integral edge detection, image recovery, neural network edge identification, etc. The algorithms are used to extract some feature information from the weld seam [24]. In the context of LWAM, the shape of the molten pool can be the main feature to be extracted and apply a control rule. In summary, using an adapted camera, the tracked information in the weld pool can be obtained by computer image processing techniques and provides real-time status of the process.

A molten pool frame image is shown in Fig. 1.

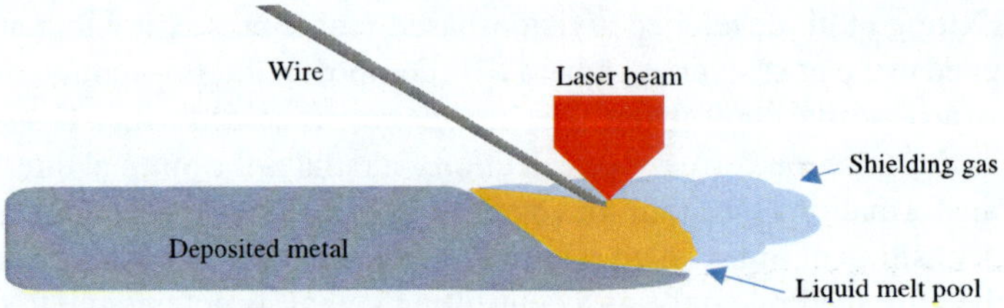

Fig. 1 A complete frame of a molten pool image

3.1 Image Data Acquisition

Real-time monitoring of weld pool dynamics depends on the extracted information used to calculate the weld pool's geometric features [25, 26]. In the image acquisition, a camera coupled with a robust image processing algorithm is crucial. Usually, the molten pool's raw image cannot be directly exploited because of the nonlinearity and disturbances in the process. These two factors add difficulties in image acquisition and processing. Image processing aims to get the relevant information by using appropriate image features while suppressing undesired distortions. However, the relative motion between the image acquisition device and the weld pool significantly affects the information acquirement.

A flow chart of a molten pool image processing is shown in Fig. 2.

The following steps describe an image extraction process of the molten pool. The algorithm was coded in Python and is composed of the following steps:

Image Filtering: First, we removed the noise in the images. During the image acquisition and quantification, interferences are present and cause some signal disturbances that lead to fuzzy images. A Gaussian and median filter [27] are combined to obtain a better image while preserving the weld pool edge. Also, considering the processing speed, a median filter gives the processed outputs faster.

Edge Detection: A Sobel filter [28] is then applied to create a frame for easy recognition of the melt pool profile. Essential features in the image are preserved and non-necessary characteristics removed. We noted a high reduction in the image's size after this step.

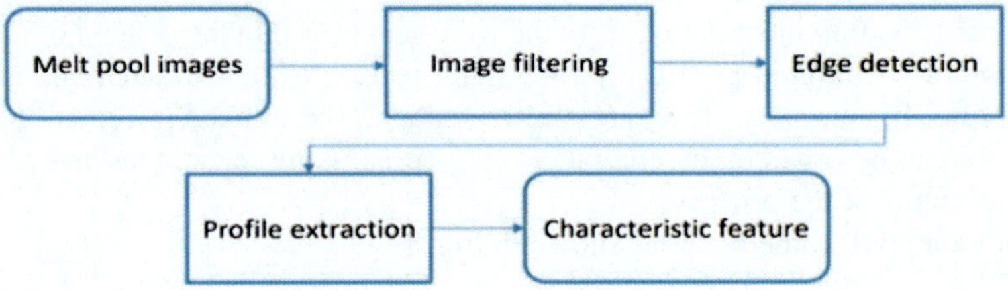

Fig. 2 Flow chart of molten pool image processing during the deposition process

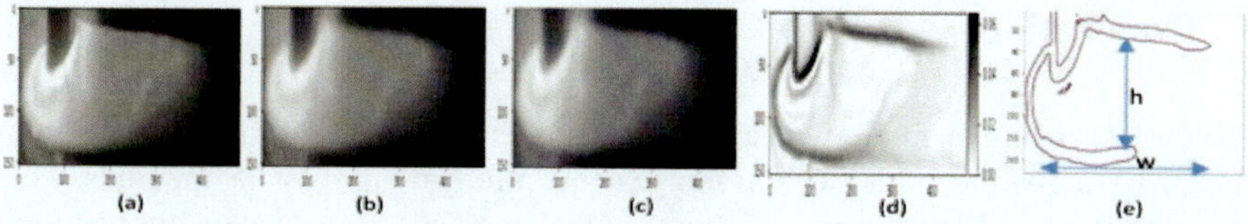

Fig. 3 **a** Original, **b** Gaussian filter, **c** Median filter, **d** Edge detecting, **e** Profile extracting

Profile Extraction: Finally, a contour extraction is applied to reconstruct the image obtained from edge detection while keeping the pool's shape precisely. Various feature extraction techniques, such as Fourier descriptor, extraction of occupancy, endpoints features, Zernike moments, etc., are proven to be efficient in more complex tasks [29].

The results of the image processing are shown in Fig. 3. The image size is 153 X 482 pixels, a standard deviation of 4 for the Gaussian filter, a Kernel size of 7 for the median filter and a contour level of 0.02 are used.

4 Experimental Set-up Approach

In robotics AM, regulating only the laser power or wire feeding velocity is not enough to obtain a stable deposition. The laser power and the wire feed rate are the most important parameters because the weld seam size increases with these parameters. An experimental set-up composed of a Laser power source, a wire feeder, a camera, a controller, and the robot is proposed in Fig. 4. The profile in Fig. 3. Can be used as a reference shape where any change in the shape profile or the width (w) or height (h) of the melt pool would generate an error. The relationship between the process errors and the molten pool's process parameters must be defined using a control rule, and finally, the control center will monitor the deposition process. The proposed set-up is shown in Fig. 4.

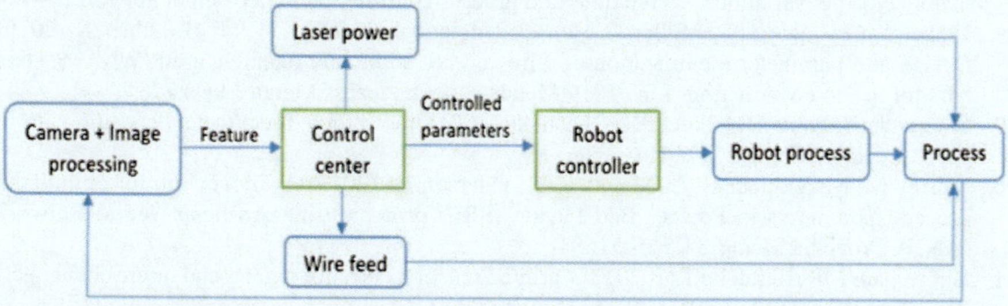

Fig. 4 Structure diagram of the proposed experimental set-up

5 Conclusion

The use of sensing technology in additive manufacturing processes has made significant progress. Many fields, such as image acquisition and processing methods, weld seam modeling, and process parameters control, are needed to develop online monitoring for AM. With the development of machine learning, researchers and engineers of AM subfields are urged to work together to create a real intelligent seam tracking of the 3D printing process to make the future of 3D printing robots brighter. This paper reviewed the literature on the use of a vision-based approach in the additive manufacturing process. A case study of the profile extraction process was also performed, and a sensing methodology to inspire future research was presented. The image processing results could be used to generate a graphical representation of the processed images and develop a control rule to monitor the deposition paths.

Acknowledgments This work was supported by the Interrreg V-A Grande Région "Fabrication Additive par Dépôt de Fil" (Fafil) project.

References

1. Gibson I, Rosen D, Stucker B (2010) Additive manufacturing technologies, vol 238. Springer
2. Yakout M, Elbestawi M, Veldhuis SC (2018) A review of metal additive manufacturing technologie. Solid State Phenom 278:1–14
3. Evjemo LD, Moe S, Gravdahl JT, Dubonnet OR, Gellein LT, Br⊘tan V (2017) Additive manufacturing by robot manipulator: an overview of the state-of-the-art and proof-of-concept results. In: 22nd IEEE ETFA
4. Fernando R (1998) 3D printing with metals. Comput Control Eng
5. Khorasani AM, Gibson I, Goldberg M, Littlefair G (2016) A survey on mechanisms and critical parameters on solidification of selective laser melting during fabrication of Ti-6Al 4V prosthetic acetabular cup. Mater Des 103(5):348–355
6. Chen SB, Wu J Real-time control of weld pool dynamics during robotic GTAW. Intellized Methodol Arc Weld Dyn Process 221–273
7. Regan P, Prickett P, Setchi R, Hankins G, Jones N (2016) Metal based additive layer manufacturing: variations, correlations and process control. Procedia Comput Sci 96:216–224
8. Müller J, Grabowski M, Müller C, Hensel, Unglaub J, Thiele K, Kloft H, Dilger K (2019) Design and parameter identification of wire and arc additively manufactured (WAAM) steel bars for use in construction. June 2019 Metals—Open Access Metall J 9(7):725
9. Mazumder J, Dutta D, Kikuchi N, Ghosh A (2000) Closed loop direct metal deposition: art to part. Opt Laser Eng 34:397–414
10. Marrey M, Malekipour E, El-Mounayri H, Faierson EJ (2019) A Framework for optimizing process parameters in Powder Bed Fusion (PBF) process using Artificial Neural Network (ANN). Procedia Manuf 34:505–515
11. Khosravani MR, Reinicke T (2020) 3D-printed sensors: current progress and future challenges. Sens Actuators A: Phys 305:111916, 15 April 2020
12. Bi G, Gasser A, Wissenbach K, Drenker A, Poprawe R (2006) Characterization of the process control for the direct laser metallic powder deposition. Surf Coat Technol 201(6):2676–2683
13. Toyserkani E, Khajepour A (2006) A mechatronics approach to laser powder deposition process. Mechatronics 16(10):631–641

14. Cheng Y, Jafari M (2008) Vision-based online process control in manufacturing applications. IEEE Trans Autom Sci Eng 5(1):140–153
15. Heralić A, Christiansson AK, Ottosson M, Lennartson B (2010) Increased stability in laser metal wire deposition through feedback from optical measurements. Opt Lasers Eng 48(4):478–485
16. Doubenskaia M, Pavlov M, Grigoriev S, Tikhonova E, Smurov I (2012) Comprehensive optical monitoring of selective laser melting. Publ 2012 Phys J Laser Micro Nanoeng
17. Xiong J, Zhang G (2013) Online measurement of bead geometry in GMAW-based additive manufacturing using passive vision. 2013 Meas Sci Technol 24:115103
18. Ocylok S, Alexeev E, Mann S, Weisheit A, Wissenbach K, Kelbassa I (2014) Correlations of melt pool geometry and process parameters during laser metal deposition by coaxial process monitoring. Phys Procedia 56:228–238
19. Gobert C, Reutzel EW, Petrich J, Nassar AR, Phoha S (2018) Application of supervised machine learning for defect detection during metallic powder bed fusion additive manufacturing using high resolution imaging. Addit Manuf 21:517–528
20. Clijsters S, Craeghs T, Buls S, Kempen K, Kruth JP (2014) In situ quality control of the selective laser melting process using a high-speed, real-time melt pool monitoring system. Publ 2014 Int J Adv Manuf Technol Vol
21. Xiong J, Liu G, Pi Y (2019) Increasing stability in robotic GTA-based additive manufacturing through optical measurement and feedback control. Robot Comput-Integr Manuf 59:385–393
22. Pan Y https://www.universal-robots.com/blog/simplify-robot-programming-with-g-code. [Online accessed: 27–08–20]
23. Nandi C, Caspi A, Grossman D, Tatlock Z (2017) Programming language tools and techniques for 3D printing. SNAPL 2017
24. Yana X, Gu P (1996) A review of rapid prototyping technologies and systems. Comput Aided Des 28(4):307–318
25. Chen SB, Wu J (2008)Intelligentized methodology for arc welding dynamical processes visual information acquiring, knowledge modeling and intelligent control. Springer, p. 72
26. Zhang W, Liu Y, Wang YX, Zhang Y (2012) Characterization of three-dimensional weld pool surface in gas tungsten arc welding. July 2012 Weld J 91(7)
27. Kumar A, Sodhi SS (2020) Comparative analysis of gaussian filter, median filter and denoise autoenocoder. In: Published in: 2020 7th international conference on computing for sustainable global development
28. Ganesan P, Sajiv G (2017) A comprehensive study of edge detection for image processing applications. In: Published in: 2017 international conference on innovations in information, embedded and communication systems (ICIIECS)
29. Kumar G, Bhatia PK (2014) A detailed review of feature extraction in image processing systems. In: February 2014 IEEE fourth international conference on advanced computing & communication technologies

Influence of Friction Stir Welding Parameters on Axial Forces and Mechanical Properties of Aluminium Alloy 7075

Check for updates

Angkarn Kamruan, Somdech Ingkawara, Kodcharawit Lamkam, and Somsak Siwadamrongpong

Abstract Friction stir welding (FSW) process was widely used in many industries due to using of aluminium was rising for light-weight design. It was generally known that some grades of aluminium have low melting weldability which might lead to cracks, porosity, warping and deformation, such as 7075 aluminum alloy. The aim of this research was to investigate the effect of FSW parameters on the axial forces and mechanical properties of Aluminium alloy 7075. The three welding parameters were studied, rotation speed, welding speed and stir tool shoulder diameter. The experiment was performed on force measuring fixture with three-axis vertical milling machine. Shoulder diameter was the only main effect that significantly affect to axial force. The minimum axial force was 3205 N with rotational speeds of 1580 rpm, welding speed of 63 mm/min and shoulder diameter 15 mm which gave tensile strength of 325 MPa or 67% of based material. On the other hand, shoulder diameter and welding speed were the main effects which significantly affect tensile strength. The maximum tensile strength was 331 MPa or 68% of base metal with rotational speeds of 1580 rpm, welding speed of 36 mm/min and shoulder diameter 9 mm.

Keywords Friction stir welding · Aluminum alloy · Al7075 · Tensile strength · Axial forces

---

A. Kamruan
Industrial Management Engineering, Faculty of Industrial Technology, Nakhon Ratchasima Rajabhat University, Nai Muang, 340 Suranarai Road, Nakhon Ratchasima, Thailand

S. Ingkawara
Department of Industrial Engineering, Faculty of Engineering, Pathumwan Institute of Technology, 833 Rama 1 Road, Wangmai, Pathumwan, Bangkok, Thailand

K. Lamkam · S. Siwadamrongpong (✉)
School of Manufacturing Engineering, Institute of Engineering, Suranaree University of Technology, 111 University Avenue, Muang, Nakhon Ratchasima, Thailand
e-mail: somsaksi@sut.ac.th

G. Conte and O. Sename (eds.), *Proceedings of 10th International Conference on Mechatronics and Control Engineering*, Lecture Notes in Mechanical Engineering,
https://doi.org/10.1007/978-981-19-1540-6_4

1 Introduction

Aluminium was widely used in many industries such as electric vehicles, high-speed trains and aircraft components. Because of its lightweight, good mechanical properties compared to weight and corrosion-resistant. However, there was generally known that some grades of aluminium were low fusion weldability and might yield crack or porosity problems after welding, such as 7075 aluminium alloy. The fusion welding also might create harmful fumes for operators and environment [1]. Therefore, friction stir welding (FSW) which was one of solid-state welding techniques was initiated for better mechanical properties in such aluminium grades. The welding of Al7075 [2] was studied using three different welding processes: MIG, TIG, and FSW and investigated on microstructure and mechanical properties. It was reported that smaller grain sizes were found with FSW compared to other welding processes. The highest tensile strength was also the test piece under FSW. Friction stir welding was yielded lower distortions and residual stresses at weldment. It was discussed as lower heat obtained from welding by FSW. However, the friction stir welding often had problems with high axial loads to plug and feed stir tool into workpiece. These also might lead to wear and tear of the machine parts [3]. Moreover, the axial force was also reported that higher axial force affects higher weld strength, in particular welding of high-strength aluminium grades, such as Al7075 aluminium [4]. According to the report, optimum parameters of FSW in Al7075 aluminium were rotation speed of 1400 rpm, welding speed of 60 mm/min, which caused an axial force of 8000 N and yielded the maximum tensile strength of 397 MPa [5]. In addition, for other grades of aluminium, the axial force was also reported that the axial force affected microstructure and mechanical properties, for instance, Al5052. The objective of this research was to study influence of FSW parameters on axial forces and mechanical properties of 7075 aluminium alloy. The parameters were rotational speed, weld speed and shoulder diameter.

2 Material and Methods

The 7075 aluminium alloy rolled plate was into workpiece dimensions of 100 × 50 × 4 mm by vertical milling machine. Chemical composition by spectrometer and tested mechanical properties were shown in Table 1. Two workpieces were set up on

Table 1 Chemical composition and Mechanical properties of Al7075

Composition (wt%)							Mechanical properties			
Mg	Mn	Zn	Fe	Cu	Si	Al	Yield Strength (MPa)	Tensile Strength (MPa)	Elongation (%)	Vicker Hardness (HV)
2.1	0.12	5.1	0.35	1.2	0.58	Bal	410	485	12	160

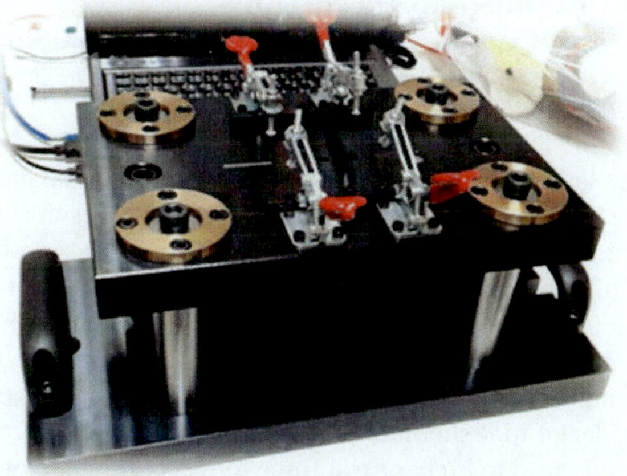

Fig. 1 Force measuring fixture [7]

force measurable fixture as indicated in Fig. 1. Stir tool with a cylindrical pin 3 mm in diameter was made by SKD61 hot-work tool steel and hardened to 51–55 HRC. Reference [6] which displayed in Fig. 2, respectively. The welding experiment was performed using 3-axis Haven XLW6332 HIV vertical milling machine. 2^k factorial design was employed to analyzed 3 FSW parameters as indicated in Table 2. FSW was carried out with fixed tool tilt angle 3°. The axial force was measured during welding. Welded sample was cut into specimen for tensile test following ASTM-E8M standard.

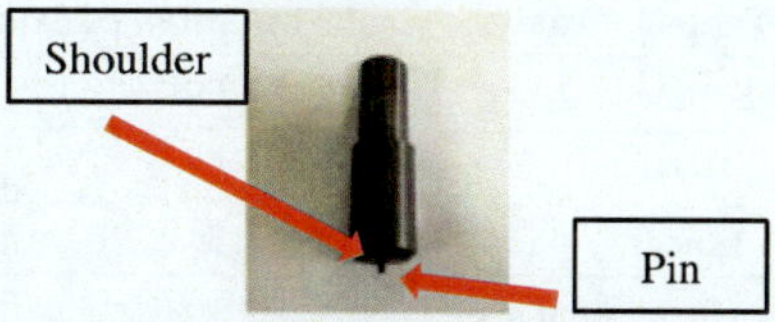

Fig. 2 Tool geometry

Table 2 FSW process parameters

Parameters	Low	high	Unit
Rotation speed	1580	2220	rpm
Welding speed	36	63	mm/min
Shoulder diameter	9	15	mm

3 Results and Discussions

3.1 Axial Force

Table 3 displayed axial force which was the maximum axial force during feed stir tool along the welding line. From analysed results of DOE showed in Table 4, tool shoulder diameter was only main effect that significant influence the axial force while the others were not significant at a confident level of 0.05. The main effect plot on axial force was illustrated in Fig. 3. The minimum axial force was 3205 N with rotational speed of 2220 rpm, welding speed 36 mm/min and shoulder diameter 9 mm which given tensile strength of 325 MPa or 67% of based material [8]. The wider of tool shoulder was discussed as higher area of bearing compression force might lead to higher axial force to reach the same pressure on workpiece.

Table 3 Experimental results

Parameters/RUN	1	2	3	4	5	6	7	8
Rotational speed (rpm)	1580	1580	1580	1580	2220	2220	2220	2220
Welding speed (mm/min)	36	36	63	63	36	36	63	63
Shoulder Diameter (mm)	9	15	9	15	9	15	9	15
Axial Force (N)	3433	5465	3825	6301	3205	4893	3900	5682
Tensile Strength (MPa)	331	269	313	251	325	268	285	241

Table 4 Analysis results of design of experiment on axial force

Source	DF	Adj SS	Adj MS	F-Value	P-Value
Model	6	9,280,273	1,546,712	101.01	0.076
Linear	3	9,101,221	3,033,740	198.12	0.052
Rotational speed (A)	1	225,792	225,792	14.75	0.162
Welding speed (B)	1	919,368	919,368	60.04	0.082
Shoulder diameter (C)	1	7,956,061	7,956,061	519.58	0.028
2-Way Interactions	3	179,053	59,684	3.90	0.353
A*B	1	8192	8192	0.53	0.598
A*C	1	134,680	134,680	8.80	0.207
B*C	1	36,181	36,181	2.36	0.367
Error	1	15,312	15,312		
Total	7	9,295,586			
Model Summary	S	R-sq	R-sq (adj)	R-sq (pred)	
	123.744	99.84%	98.85%	89.46%	

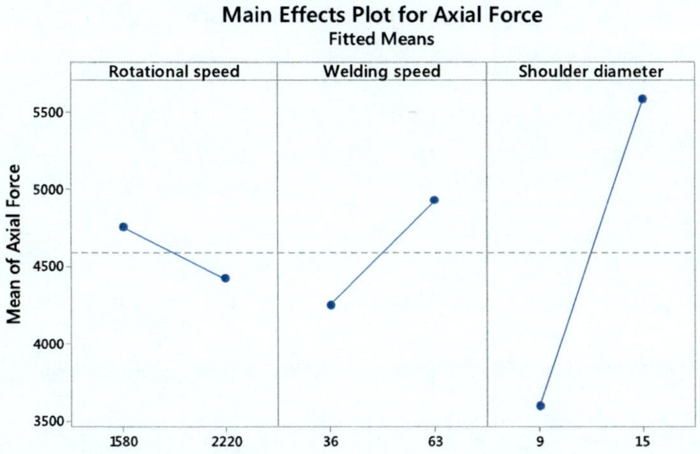

Fig. 3 Main effects plot for axial force

3.2 *Tensile Strength*

Table 3 illustrated the average tensile strength from 3 specimen. Table 5 indicated analysed results of DOE, it was found that shoulder diameter and welding speed were the main effects which significantly influence the tensile strength while rotational speed was not significant at the test range with a confident level of 0.05. The main effect plot on tensile strength was indicated in Fig. 4. The interaction between rota-

Table 5 Analysis results of design of experiment on tensile strength

Source	DF	Adj SS	Adj MS	F-Value	P-Value
Model	7	24,948.0	3564.0	14.67	0.000
Linear	3	22,189.5	7396.5	30.45	0.000
Rotational speed (A)	1	63.1	63.1	0.26	0.617
Welding speed (B)	1	5465.3	5465.3	22.50	0.000
Shoulder diameter (C)	1	16,661.1	16,661.1	68.59	0.000
2-Way Interactions	3	2737.2	912.4	3.76	0.032
A*B	1	2360.4	2360.4	9.72	0.007
A*C	1	41.5	41.5	0.17	0.685
B*C	1	335.3	335.3	1.38	0.257
3-Way Interactions	1	21.3	21.3	0.09	0.771
Error	16	3886.3	242.9		
Total	23	28,834.3			
Model Summary	S	R-sq	R-sq (adj)	R-sq (pred)	
	15.5851	86.52%	80.63%	69.67%	

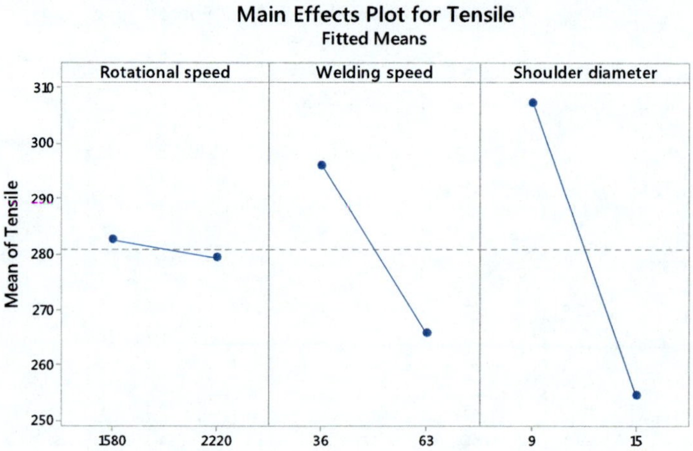

Fig. 4 Main effects plot for tensile

tional speed and welding speed was also a significant affect on the tensile strength. It could be discussed that a bigger shoulder diameter may result in high heat generation [9] and consequently lead to affect the microstructure within the heat affect zone (HAZ) area [10]. The grain growth in HAZ area might reflect in decrease of tensile strength. The maximum tensile strength was 331 MPa or 68% of base metal at rotational speeds of 1580 rpm, welding speed 36 mm/min and shoulder diameter 9 mm which was acceptable strength of welding work [11].

4 Conclusions

3 FSW parameters including rotational speed, welding speed and tool shoulder diameter were investigated on axial force and tensile strength of 7075 aluminium alloy. Analysis of DOE results were revealed that;

- Shoulder diameter was the only main effect that significant affect to axial force. The minimum axial force was 3205 N with rotational speeds of 2220 rpm, welding speed of 36 mm/min and shoulder diameter 9 mm which gives tensile strength of 325 MPa or 67% of based material.
- Shoulder diameter and welding speed were the main effect which significant affect tensile strength. The maximum tensile strength was 331 MPa or 68% of base metal with rotational speeds of 1580 rpm, welding speed of 36 mm/min and shoulder diameter 9 mm.

Acknowledgements The authors would like to thank you Suranaree University of Technology for research grant and support in equipment.

References

1. Baffari D, Buffa G, Campanella D, Fratini L, Micari F (2014) Friction based solid state welding techniques for transportation industry applications. Procedia CIRP 18:162–167
2. Yeni C, Sayer SAMİ, Pakdil M (2009) Comparison of mechanical and microstructural behaviour of TIG, MIG and friction stir welded 7075 aluminium alloy. Kovove Mater 47:341–347
3. Elangovan K, Balasubramanian V, Valliappan M (2008) Influences of tool pin profile and axial force on the formation of friction stir processing zone in AA6061 aluminium alloy. Int J Adv Manuf Technol 38(3–4):285–295
4. Rajakumar S, Muralidharan C, Balasubramanian V (2011) Influence of friction stir welding process and tool parameters on strength properties of AA7075-T6 aluminium alloy joints. Mater Des 32(2):535–549
5. Moshwan R, Yusof F, Hassan MA, Rahmat SM (2015) Effect of tool rotational speed on force generation, microstructure and mechanical properties of friction stir welded Al–Mg–Cr–Mn (AA 5052-O) alloy. Mater Des (1980–2015), 66:118–128
6. Kasman Ş, Yenier Z (2014) Analyzing dissimilar friction stir welding of AA5754/AA7075. Int J Adv Manuf Technol 70(1–4):145–156
7. Kamruan A, Siwadamrongpong S, Thanasubtawee U, Muangjunburee P (2020) Influence of friction stir welding parameters on tensile strength of semi-solid cast 2024 aluminum alloy butt joints. J KMUTNB 3:2465–4698
8. Jassim AH, Abtan NS, Marmoos MS (2019) Study on the effects of rotational and transverse speed on temperature distribution through friction stir welding of AA2024-T3 aluminium alloy. J Adv Res Fluid Mech Therm Sci 53(2):234–248
9. Ramanjaneyulu K, Madhusudhan Reddy G, Venugopal Rao A (2014) Role of tool shoulder diameter in friction stir welding: an analysis of the temperature and plastic deformation of AA 2014 aluminium alloy. Trans Indian Inst Metals 67(5):769–780
10. Dawood HI, Mohammed KS, Rahmat A, Uday MB (2015) Effect of small tool pin profiles on microstructures and mechanical properties of 6061 aluminum alloy by friction stir welding 25(9):2856–2865
11. Prakash P, Jha SK, Lal SP (2013) A study of process parameters of friction stir welded AA 6061 aluminum alloy. Int J Innov Res Sci, Eng Technol 2(6):2304–2309

Mechanical Properties of Weld Joint by Friction Stir Welding on Aluminum Alloy 5083

Siriporn Khantongkum, Ukrit Thanasubtawee, Winut Promdan, and Somsak Siwadamrongpong

Abstract The manufacturing industry, aluminum alloys were used in various applications such as the electric vehicle industry, marine industry, and also aircraft industry. However, welding on some grades of aluminum alloys, for instant Al5083, might lead to problems at weldment such as crack, pinholes, or incomplete weld joint. Friction stir welding was one of solid-state welding which might suit for such aluminum alloys. Therefore, this research was aimed to investigate mechanical properties of weldment on aluminum alloy 5083 welded by friction stir welding. Small friction stir welding machine was applied in this study. Two-factor of the welding parameters, rotation speed, and welding speed, were studied. The rotation speed of three levels (1200, 1500, 1800 rpm), and two levels of welding speed (30 and 60 mm/min) were employed. It was found that rotation speed of 1800 rpm and welding speed of 30 mm/min was found to yield good weldment and maximum tensile strength 224.41 MPa or 68.65% of based material.

Keywords Friction stir welding · Stir tool · Al5083 · Aluminum alloys

1 Introduction

In manufacturing processes, metal welding is the process of joining metal materials. The conventional welding techniques were fusion state welding which part of workpieces were melted together and additive metal might be added as an additive

S. Khantongkum
Department of Production Engineering, Faculty of Engineering, Chaiyaphum Rajabhat University, 167, Muang Chaiyaphum 36000, Thailand

U. Thanasubtawee
Department of Industrial Engineering, Faculty of Engineering, Rajamangala University of Technology Lanna Tak, 41/1 Paholayothin road, Mai Ngam, Muang Tak 63000, Thailand

W. Promdan · S. Siwadamrongpong (✉)
School of Manufacturing Engineering, Institute of Engineering, Suranaree Sub-District, Suranaree University of Technology, 111 University Avenue, Muang, Nakhon Ratchasima 30000, Thailand
e-mail: somsaksi@sut.ac.th

G. Conte and O. Sename (eds.), *Proceedings of 10th International Conference on Mechatronics and Control Engineering*, Lecture Notes in Mechanical Engineering,
https://doi.org/10.1007/978-981-19-1540-6_5

to the welded zone. During the melt state, the welded zone will be covered with gases to prevent oxidation of the metal and then the weld joint was cooled down and strengthened. There are many types of metal welding such as metal inert gas welding, resistance welding, and laser welding. However, the fusion techniques usually used electrical energy as an energy source and might release emissions to the atmosphere. The air pollution generated from the industrial sector such as welding fumes and toxic gases was under consideration. In fumes from welding process, including of mixture of metal oxide, chemical complex of metal oxides, silicate, and fluoride. [1, 2]. Recently, problems of pollution from small particles less than 2.5 μ (PM2.5) accumulated in urban areas were rising. Thailand had an estimated 37,500 people who die prematurely as a result of these PM2.5 problems [3].

The problem of welding metal and small dust particles PM2.5, therefore, non-polluting metal welding was invented. One of the solid-state welding techniques, friction stir welding method (FSW), was first found to replace some metals that was hard-to-weld with the fusion technique, especially for aluminum. The advantages included the microstructure of fines grains and good weld metallurgical properties with low emission of greenhouse gasses [4]. The current trend and situation of automotive industry to reduce product weight that might lead to low resource consumption (raw materials, fuel, or electricity). There was forecasted that the future used material in automotive industry might change to the use aluminum alloy in electric cars. [5] The research of Thomas et al. [6] was indicated that the friction stir welding process could be applied in the automotive industry.

The friction stir process parameters were consisted of factors that affects welding performance such as geometry of stir tool, tool rotation speed, and welding speed. These factors affected the mechanical properties of the weldment as reported by Trimble et al. [7], Abdollahzadeh et al. [8], Kuntala Goutham Krishna et al. [9]. The Al5083 was one grade of aluminum that was known as hard-to-weld with fusion state technique. Therefore, the development of friction stir welding for such aluminium could create knowledge for further industrial applications. This research paper was aimed to investigate the effects of friction stir welding parameters on 5083 aluminum alloys.

2 Method and Material

Aluminum alloy 5083 plate was prepared for friction stir welding, sized 100 × 50 *mm with 4 mm thickness. The chemical composition of the Al5083 was indicated in* Table 1. *2 plates were set up on the welding jig. The pin of stir tool was cylindrical in*

Table 1 Chemical composition of Al5083 (%) by spectrometer

Mg	Mn	Fe	Si	Cr	Cu	Ti	Al
4.22	0.51	0.31	0.12	0.07	0.02	0.02	Bal

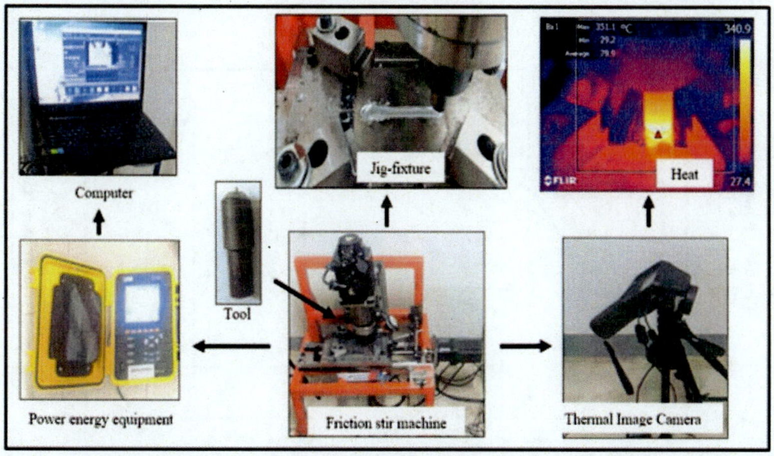

Fig. 1 Set up of friction stir welding and measuring devices

Table 2 Welding parameters

Rotation speed	1,200 rpm	1,500 rpm	1,800 rpm
Welding speed	30 mm/min	60 mm/min	–
Tool/Pin shoulder diameter	15 mm/3 mm		

shape with 3 mm in diameter and tool shoulder diameter was 15 mm as indicated in Fig. 1. *The welding parameters were illustrated in* Table 2. *Full factorial design was employed to analyze the influence of two welding parameters, rotation speed, and welding speed. Tool shape and tilt angle parameters were fix as displayed in* Table 2. *The welded workpiece was cut into specimen shape for tensile test following ASTM standard. The specimen preparation was done with CNC milling machine.* Figure 1 *was showed the set-up of friction stir welding configuration on a small friction stir welding machine. Thermal imaging cameras and electrical analysis devices were also set-up for temperature and power using data along the welding. The stir pin was plugged into the plate at desired rotation speed until the tool shoulder was touched the plate as the electrical power was rapidly raised, as shown in* Fig. 2. *Then, the stir tool was dwelled for 5 min for rising of temperature and start to weld at specified welding speed.*

3 Result

From the welding samples, it was found that two plates were joined together along the weld line. The weldment width was wide as shoulder size, 15 mm, and some fin was found along the edge of weldment as displayed in Fig. 3. The similar results

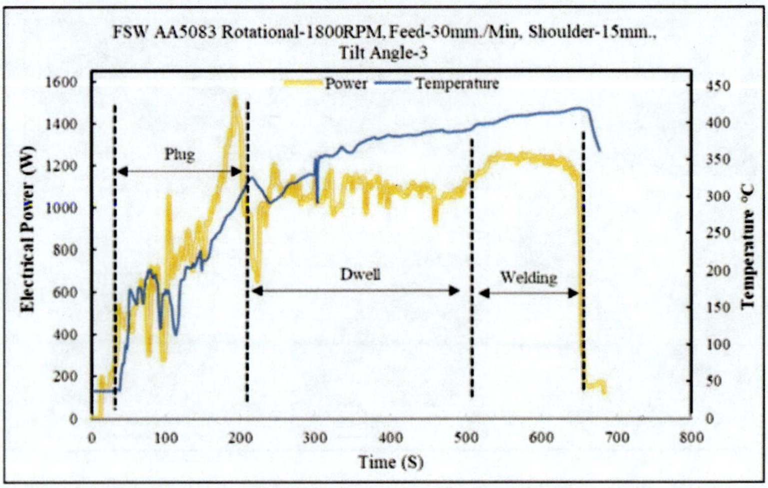

Fig. 2 Pattern of electrical power and temperature during welding

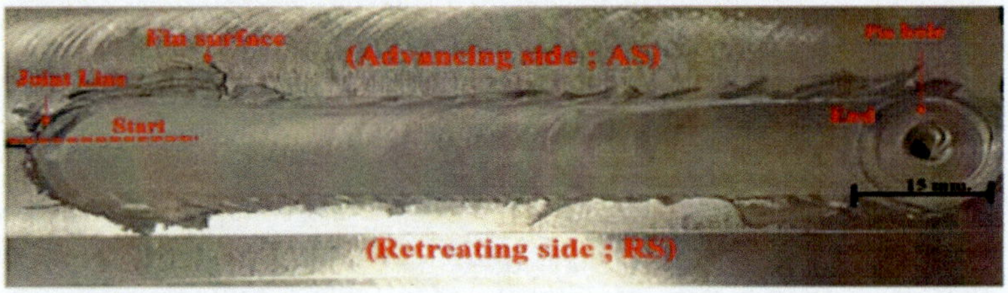

Fig. 3 Welded Surface of Specimen after Friction stir welding

also reported by Surasit Rawangwong et al. [10] that friction stir welding surface was glittering and circular mark of the flow of material on the weldment.

From the tensile strength results, it was found that rotational speed of 1,800 rpm and welding speed of 30 mm/min was yielded the highest strength of 224.41 MPa or 68.65% of based material and illustrated in Fig. 4. Table 3 showed the analysis results of DOE on the tensile strength of specimen. It was found that interaction between rotational speed and welding speed was the most important parameter to tensile strength. Main effects, rotational speed, and welding speed, also significantly influence the tensile strength with significant level of 0.05. It might be considered that higher rotational speed help in rising of temperature and lead to higher strength. However, the welding speed of 60 mm/min and rotational speed of 1800 rpm was yielded the lower of strength, it could be considered that the too high welding speed may make incomplete of material flow at the edge of stir zone that may lead to micro void, Huijie Liu et al. [11].

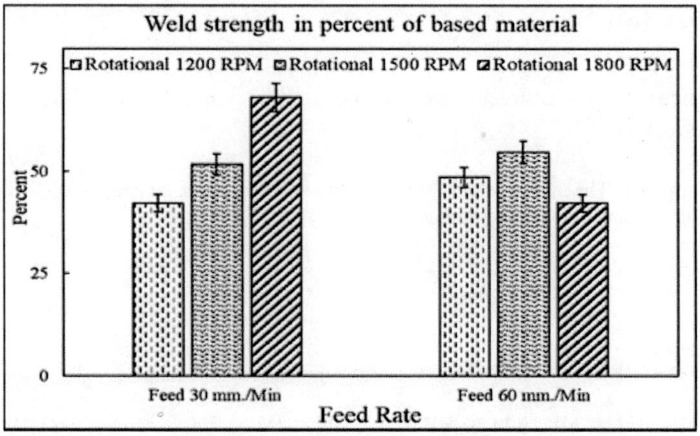

Fig. 4 Tensile strength of specimen

Table 3 Analysis of design of experiment results on tensile strength

Source	DF	Adj SS	Adj MS	F-Value	P-Value
Model	5	10,337.1	2067.42	25.57	0.001
Linear	3	3543.8	1181.27	14.61	0.004
Rotational speed	2	2532.2	1266.09	15.66	0.004
Feed rate	1	1011.6	1011.64	12.51	0.012
2-Way interactions	2	6793.7	3396.64	42.01	0.000
Rotational speed * feed rate	2	6793.7	3396.64	42.01	0.000
Error	6	485.1	80.85		
Total	11	10,822.2			

Model summary	S	R-sq	R-sq (adj)	R-sq (pred)
	8.99172	95.52%	91.78%	82.07%

4 Conclusion

The weld joint of Al5083 was carried out by friction stir welding. All welding conditions were in good surface with fin along the edge of weldment. The welding parameters were found that interaction between rotational speed and welding speed was the most important parameter to tensile strength. Main effects, rotational speed, and welding speed, also significantly influence the tensile strength with significant level of 0.05. The highest tensile strength was found with rotational speed of 1,800 rpm and welding speed of 30 mm/min at 224.41 MPa or 68.65% of based material.

5 Suggestion

The influence of temperature and electrical power to tensile strength should be studied in further work.

Acknowledgments The authors would like to thank you Suranaree University of Technology for the research grant and support in equipment.

References

1. Ranaa HK, Akhtar MR, Ahmed MB, Liò P, Quinn JMW, Huq F, Monie MA (2019) Genetic effects of welding fumes on the progression of neurodegenerative diseases. Neuro Toxicol 71:93–101
2. Sansuksom P (2019) Risk assessment of worker exposure to metal fume produced in welding process. J Suvarnabhumi Inst Technol
3. The annual State of Global Air.org (2021) "Average Annual Population-Weighted PM2.5" data. Accessed 30 June 2021
4. Ma RSMAZY (2005) Friction stir welding and processing. Mater Sci Eng 50: 1–78
5. Orłowicz AW, Mróz M, Tupaj M, Trytek A (2015) Materials used in the automotive industry. Archives of foundry engineering, vol 15. ISSN (2299–2944)
6. Thomas WM, Kallee S, Staines DG, Oakley PJ (2006) Friction stir welding—process variants and developments in the automotive industry. In: SAE World Congress
7. Trimble D, O'Donnell GE, Monaghan J (2015) Characterisation of tool shape and rotational speed for increased speed during friction stir welding of AA2024-T3. J Manuf Process 15, 495–500, 17:141–150
8. Abdollahzadeh A, Shokuhfar A, Cabrera JM, Zhilyaev AP, Omidvar H (2019) In-situ nanocomposite in friction stir welding of 6061-T6 aluminum alloy to AZ31 magnesium allo. J Mater Process Tech 263: 296–307
9. Kuntala Goutham Krishna AD, Manichandra B (2017) Study on mechanical propreties of friction stir welded dissimilar AA2024 and AA7075 aluminum alloy joints. Int J Nanotechnol Appl 11:285–291
10. Rawangwong S, Boonchouytan W, Homkhiew C, Cheewawuttipong W (2015) Effect of parameter in friction stir welding dissimilar joints of aluminum semi-solid metal between SSM 356 and 7075. Naresuan Univ Eng J 21–30
11. Liu H, Zhang H, Pan Q, Yu L (2012) Effect of friction stir welding parameters on microstructural characteristics and mechanical properties of 2219-T6 aluminum alloy joints. Int J Mater Form 235–241

A Prototype NIRS Device to Increase Safety of Diving

Davide Animobono, David Scaradozzi, and Giuseppe Conte

Abstract A simple prototype device that may become part of the standard divers' safety equipment is designed, constructed, and tested in order to prove the practicability of the underlying idea. The device is based on near-infrared spectroscopy (NIRS) and can detect drops in oxyhemoglobin and rises in deoxyhemoglobin concentrations in the blood. The proposed device not only collects biometric data, but can also autonomously identify and promptly notify anomalies so as to warn and speed up appropriate rescue activities. Easiness of use and low costs are important requirements in order to make the device suitable for everyday use by most divers. The first demonstration tests, carried out in laboratory conditions, are illustrated and discussed.

Keywords Near-infrared spectroscopy · Biometric device · Feature extraction · Diving safety

1 Introduction

Underwater diving is an important activity, motivated by professional, scientific, and recreational purposes. In the last decades, diving technologies have greatly progressed, but significant risks are still present [5]. Wearable, biometric devices capable to monitor the physiological condition of the diver can increase the safety and effectiveness of traditional practices. Moreover, collected data can support and foster scientific studies on the physiology of diving.

D. Animobono (✉) · D. Scaradozzi · G. Conte
Dipartimento di Ingegneria dell'Informazione, Universitá Politecnica delle Marche, Ancona, Italy
e-mail: d.animobono@pm.univpm.it

D. Scaradozzi
e-mail: d.scaradozzi@univpm.it

G. Conte
e-mail: gconte@univpm.it

G. Conte and O. Sename (eds.), *Proceedings of 10th International Conference on Mechatronics and Control Engineering*, Lecture Notes in Mechanical Engineering,
https://doi.org/10.1007/978-981-19-1540-6_6

The near-infrared spectroscopy, or NIRS, is a non-invasive technique that allows to measure variations in oxygenated and deoxygenated hemoglobin concentrations in a biological tissue using light radiation [13]. NIRS is a well-established technology in several medical practices, for the monitoring of brain [23, 24, 26] and peripheral organs oxygenation [3, 11, 15]. The devices commonly used in the medical field have the ability to estimate the oxygen saturation [17, 23, 27], but have considerable dimensions, complexity, and costs. The possibility of measuring the oxygen saturation is also provided by pulse oximeters, which rely on the heart pulse signal, and are used only for measurements in peripheral organs.

NIRS is adequate to detect apnea-dependent hypoxia [9], but only a few studies have investigated possible underwater applications. Custom underwater NIRS devices are described in [12, 20, 29] and their use to investigate cardiovascular phenomena occurring in seals and emperor penguins during free diving is illustrated. A few other studies consider suitable NIRS devices for human swimmers and divers. In [14], a commercially available NIRS device (PortaMon by Artinis Medical Systems), placed inside a modified, commercially available, underwater silicone covering for electronic devices, is used to collect biometric data of swimmers and triathletes. Tests are carried out to measure the error introduced by the covering interposed between the device and the tissue. An underwater NIRS device that has some of the characteristics required for the usage we propose is presented in [25]. In that work, the device used in [20] is further modified to make it suitable for brain saturation measurements on elite freedivers. The result is a wearable, simple-to-use, and 100m depth-tested device capable to collect NIRS data from professional divers during their performances. All the mentioned devices are only meant to acquire and store raw signals, whose analysis is carried out offline, after returning to the surface.

The aim of this work is to design, construct, and perform preliminary tests on a prototype that overcomes some limitations of the existing devices. The novelty, with respect to underwater devices described in the literature, is the ability to process data internally, during the acquisition, so as to detect promptly a number of situations that are considered to be critical or worth signaling. This feature requires the execution of suitable data processing and data interpretation routines by an embedded logical unit. Moreover, the device needs to be able to communicate externally the results of the data processing and interpretation activities, so that the information can be passed on to the user and possibly transmitted to his companions underwater and on the surface. The economic factor is also to be considered since the proposed device should be inexpensive and thus practically usable by most divers. Only relative variations of chromophore concentrations are estimated and the identification of critical events is based on these signals. This approach is well-known in the field of Brain-Computer Interfaces (BCI) [2, 6, 7, 21, 22], where similar techniques are used to detect the activation of specific areas of the brain. This choice makes it possible to use inexpensive components and to develop simple and very general algorithms, avoiding the necessity of complex assumptions about the condition of use.

The paper is organized as follows. The structure of the device is described in detail in Sect. 2. The data processing and interpretation routines and the details of their implementation are described in Sect. 3. Preliminary tests are illustrated and discussed in Sect. 4. The last section contains the conclusions and a description of the future phases of work.

2 Structure of the Device

The device consists of two components that are connected by a short cable. The biggest component is a sealed transparent PVC cylindrical case (9 cm diameter, 19.8 cm height) that hosts the control unit of the device and a battery. The smaller component is a probe (7 cm × 4.2 cm × 3.5 cm) consisting of a sealed opaque plastic case that hosts the emitting LEDs and the receiver photodiode. Figure 1 shows the lower side of the probe with the arrangement of the two windows for the emitting LEDs on the right and of the window for the receiving photodiode on the left. The cylindrical case can be attached to the diver's belt and the probe can be fixed under the diving suit in contact with the part of the body to monitor (e.g. a muscular complex or a region of the head).

The control unit consists of a Raspberry Pi 3 model B+ board, equipped with status indicator LEDs and a digital clock with an independent power supply. A reed switch is used to turn on and off the device avoiding through-hole switches that could cause water infiltration. A rechargeable 5V power bank supplies the energy to the device. One end of the cylindrical case is equipped with an underwater cable connector (Bulgin PXP4013/10P/PC) that can be alternatively used, with different cables, to connect the probe to the control unit or to recharge the power bank.

The emitting circuit in the probe is equipped with two LEDs (Osram SFH 4860 and SFH 4850 E7800) with wavelengths $\lambda_1 = 660$ and $\lambda_2 = 850$ nm and nominal powers 3 and 50 mW, respectively. Two functionally independent circuits are present on the light-emitter board that controls the two LEDs. Each circuit is on/off controlled by a digital input and provides a constant current to one LED. A low-dropout linear voltage regulator (Analog Devices ADP667) is used in a suitable configuration to achieve this result. The receiver is an inexpensive monolithic photodiode with a single-supply

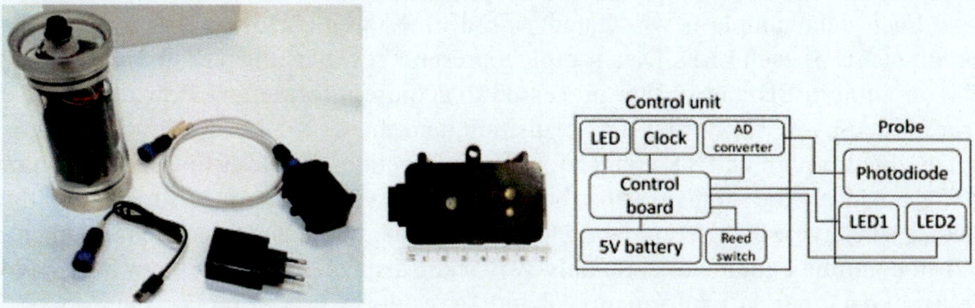

Fig. 1 The final prototype and its accessories (the lower face of the probe is shown on the right); a scheme of the device components

transimpendance amplifier (Texas Instruments OPT101). The advantages of using this device over possible alternatives are described in [18]. A Waveshare Electronics High Precision AD/DA Board that incorporates an analog-to-digital converter Texas Instruments ADS1256 acts as converting interface between the photodiode and the control board. Both the emitting and the receiving circuits are printed as PCB boards.

3 Data Acquisition, Processing, and Interpretation

In order to inspect the variations of concentration of the oxygenated hemoglobin and of the deoxygenated hemoglobin in a part of the body, like, e.g. a muscular complex or a region of the head, the lower side of the probe is placed in contact with the external area of the region. Then, the device operates by emitting infrared light that, after entering the region, is partially absorbed and partially reflected back, where it is sensed by the photodiode. The oxygenated hemoglobin and the deoxygenated hemoglobin interact differently with the infrared radiation at the two wavelengths, and the amount of the reflected radiation depends on their concentrations, denoted, respectively, as $HbO(t)$ and $HbR(t)$.

3.1 Data Acquisition

In order to generate the signal to be acquired, each LED is independently activated for a short time in an alternating sequence, and activation intervals are separated by a shut-off time. The LEDs are controlled through a suitable circuit by the control board. The durations of the activation intervals for the two wavelengths λ_1 and λ_2 and of the shut-off interval are specified in the control software implemented by the control board and, in the configuration of the device that has been used in the validation tests, they are respectively of 12 and 11 ms.

The reflected radiation generates, through the photodiode and a suitable circuit, an analog voltage signal that is processed by an AD converter that generates 7500 sps. Each data sample is associated with a timestamp and flags that indicate the on/off status of each LED. Data sample represents row data that are stored in a CSV file on a micro-SD and further processed to acquire information. Row data can be transmitted by Wi-Fi during the acquisition, and the CSV file, after its creation, is accessible through an FTP connection. By connecting a PC, Wi-Fi transmission of row data during the acquisition has been used in the validation process to analyze the device's behavior in different laboratory conditions. In the underwater environment, Wi-Fi communication works on only very short distances and it will not be used to transmit data, but only information about the results of further processing.

3.2 Signal Processing

The row data collected during the activation time of each LED and those collected during the subsequently shut-off period, excluding those obtained during the first 5 ms after each change of status, are separately averaged and the difference between the averaged values is the voltage value associated to each LED for the considered activation cycle. Discarding the row values obtained during the first 5 ms serves to eliminate the effect of transient behaviors in the LED operation. The average operation is intended to filter random measurement errors and the subtraction operation is intended to filter constant environmental radiation that possibly superimposes to the signal.

Voltage values so obtained are further averaged, separately for each LED, over a window of 10 samples to reduce high-frequency noise. This operation provides one voltage value for each LED for each time period of 460 ms (equal to the duration of 10 activation cycles of 12 ms + 11 ms for each one of the LEDs). This set of values is finally averaged using a moving window of 51 samples to reduce medium-frequency noise. This operation introduces a delay of $(460\,\text{ms} \times 51)/2 = 11,73\,\text{s}$ with respect to the acquisition time in the set of filtered data. The effects of this delay in interpreting the data will be discussed in Sect. 3.3.

Filtered data are then used to estimate the variations of the oxygenated hemoglobin concentration $\Delta_{HbO}(t)$ and of the deoxygenated hemoglobin concentration $\Delta_{HbR}(t)$ using the Modified Beer-Lambert Law [8], assuming that the emission power and the photodiode responsivity are constant and using an approximate differential pathlength factor (DPF) equal to 4. Note also that the available data do not allow the computation of the absolute concentration, but only of their relative variations. The assumption of a constant DPF could introduce some inaccuracies in the measured variations of concentrations [4, 16], and the use of a more accurate estimation could be taken into account as a future improvement. A different choice for the value of the, supposed constant, DPF can introduce a scale factor in the computed variations of concentration, but, as we will see, this does not affect the interpretation of the results.

3.3 Feature Extraction and Data Interpretation

In interpreting the data, one is interested in detecting a rise of the deoxygenated hemoglobin concentration associated with a simultaneous descent of the oxygenated hemoglobin concentration that can correspond to ischemia in the monitored region. We have designed an empirical feature-extraction algorithm assuming that the user stays at rest for about 30 s after activating the device. Values computed during the first 5 s are dropped to limit the effect of transient behavior in the action of the moving average filter. The subsequent time interval [5 s, 30 s] is used to compute the standard deviation of the deoxygenated hemoglobin concentration, namely $std_{[5,30]}(\Delta_{HbR})$. Then, we compute the following variable quantities:

1. $R_{min}(t) = min_{[30,t]}\Delta_{HbR}(t)$; we indicate the rightmost point where the minimum is attained as $\bar{t}$;
2. $O_{max}(t) = max_{[\bar{t},t]}\Delta_{HbO}(t)$;
3. $\Delta_f HbR(t) = \Delta_{HbR}(t) - R_{min}(t)$;
4. $\Delta_f HbO(t) = \Delta_{HbO}(t) - O_{max}(t)$;
5. $\Delta_{min}(t) = min\{\Delta_f HbR(t), -\Delta_f HbO(t)\}$;
6. the index $f(t) = \begin{cases} 0 & \text{for } t < 30 \\ \frac{\Delta_{min}(t)}{std_{[5,30]}(HbR)} & \text{for } t \geq 30 \end{cases}$.

The idea behind the definition of the index $f(t)$ is that of comparing increments of the deoxygenated hemoglobin concentration or decrements of the oxygenated hemoglobin, whichever is the smallest, with the standard deviation of the underling phenomenon. In computing the standard deviation, we have considered the behavior of the concentration of the deoxygenated hemoglobin because it seems to be less affected by external factors (like, e.g. the heart rate or the physiological reaction to variations of the external temperature) than the behavior of the concentration of the oxygenated hemoglobin in resting conditions. Note that $f(t)$ is not modified by adding a constant value to $\Delta_{HbR}(t)$ and $\Delta_{HbO}(t)$, nor by multiplying both of them for a constant. Then, in particular, it is invariant with respect to the choice of the DPF. The index $f(t)$ is then passed to a simple classifier that compares it with a given thresholds, whose value is chosen empirically, and that, in case the threshold is exceeded, checks if $f(t)$ stays over it for a given time.

The fact that $f(t)$ exceeds a suitably chosen threshold over a sufficiently long time interval is interpreted as the occurrence of a critical event that may indicate the establishment of ischemia in the monitored region. Clearly, the choice of the threshold and of the length of the interval over which it is exceeded is of the greatest importance for assuring a prompt detection of critical events while avoiding, at the same time, the occurrence of false alarms. At this stage, the main objective of the testing activity is to validate the idea that is behind the data interpretation procedure by showing that the behavior of $f(t)$ allows the choice of a suitable threshold and of a suitable interval over which it is exceeded in case ischemia is artificially induced (for a short time).

Note that the data interpretation procedure suffers from the delay caused by the data processing procedure described in Sect. 3.2. Moreover, a further delay, depending on the value of the threshold itself, may intervene before $f(t)$, following the trends of $\Delta_{HbR}(t)$ and of $\Delta_{HbO}(t)$, exceeds the threshold and it stays over it for a sufficiently long time. Acceptable performances imply that the global delay is less than the time required for ischemia to have dangerous effects.

The device can be configured in such a way to send a warning by Wi-Fi if the threshold is exceeded over a sufficiently long interval of time. In the underwater environment, the warning can be received by other apparatuses that are in contact or at a very close distance (less than 10 cm) from the case that hosts the device's control board. These can then use acoustic communication to transmit the warning to a surface supply vessel or take other suitable actions.

4 Test and Validation

A specific qualitative test has been designed in order to check and evaluate in laboratory conditions the ability of the device to monitor the variations of the oxygenated hemoglobin and of the deoxygenated hemoglobin and the possible establishment of ischemia. The protocol is based on the well-established Vascular Occlusion Test (VOT) [10, 19, 30], often used as a first in-vivo validation of NIRS prototypes (see, for instance, [1, 28]).

The test consists in placing the lower side of the probe on the forearm of a subject above the brachioradialis muscle and to cause ischemia in the muscle by inflating the cuff of a manual sphygmomanometer around the upper arm. The test procedure starts by measuring the subject's systolic pressure by means of the sphygmomanometer. After the measurement, the subject remains at rest for 120 s with the deflated cuff in place. Then, the NIRS probe is placed on the forearm, and data are collected for 120 s while the subject is still at rest. At the end of this phase, the cuff is inflated to a pressure 40 mmHg higher than the systolic pressure of the subject, causing the occlusion of both the venous and the arterial circulation in the arm and ischemia in the brachioradialis muscle. After 120 s the cuff is deflated, so as to restore blood circulation, and data collection continues for an additional 120 s. The total duration of the test, excluding the preliminary phase to acquire the systolic pressure of the subject, is 360 s.

The basic functioning of the device, which simply consists in measuring variations of concentrations and in recording the data, is expected to give evidence of a large decrease in the concentration of the oxygenated hemoglobin and of a simultaneous large increment in the concentration of the deoxygenated hemoglobin right after the occlusion and of the reverse phenomenon in correspondence of the restoration of blood circulation in the arm.

Concerning the detection of the ischemia, the data interpretation procedure is required to signal the phenomenon within 50 s from the establishment of the occlusion, before negative effects on the muscular complex can arise, and to avoid false alarms during the initial rest phase. The second rest phase is useful to better understand how the concentrations vary after the occlusion is removed, but no requirements are imposed in this phase to the data interpretation procedure, which is strongly influenced by the transient behavior of $\Delta_{HbR}(t)$ and $\Delta_{HbO}(t)$. As already mentioned in Sect. 3.3, in order to obtain the desired performances of the interpretation procedure, it is necessary to tune it by choosing empirically a suitable threshold and an interval over which the threshold must be exceeded.

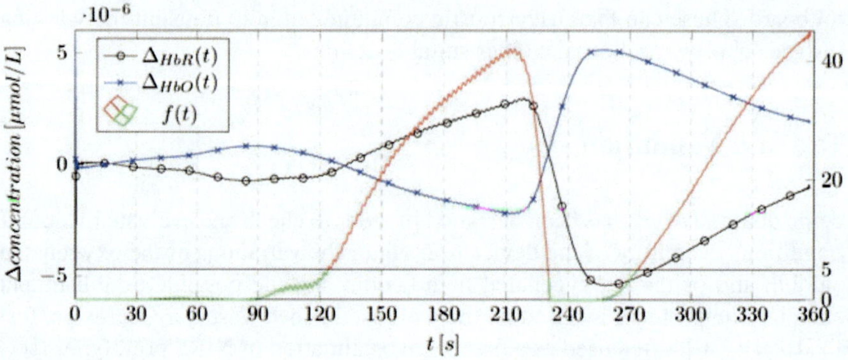

Fig. 2 Results of the test 1 in dry conditions

4.1 Tests in Dry Conditions

A number of tests have been carried out in dry conditions in the laboratory. For sake of illustration, Figs. 2, 3, and 4 show the behavior of $\Delta_{HbR}(t)$, $\Delta_{Hbo}(t)$, and $f(t)$ in three different laboratory tests (referred to as test 1, 2, and 3) in dry conditions with different subjects. Note that the time reported on the x-axis represents the instant to which the samples refer, but the values of $\Delta_{HbR}(t)$, $\Delta_{Hbo}(t)$ are actually calculated with a delay of about 12 s, as mentioned in Sect. 3.2.

Analysis of a number of data sets suggests to fix the threshold at 5 on the scale that measures $f(t)$ and at 5 s the length of the interval over which the threshold must be exceeded. The plot of $f(t)$ is drawn in red for the part exceeding the threshold.

The establishment of a critical situation due to the occlusion is correctly detected by the data interpretation procedure, i.e. $f(t)$ exceeds the threshold as expected and it stays over it for a suitable amount of time, in all the three shown data sets. In the worst case of test 3 (Fig. 4), $f(t)$ exceeds the threshold 37 s(= 25 s + 12 s) after the inflation of the cuff and stay over it for the subsequent 5 s. The total delay in detecting the critical situation is, therefore, 42 s. No false alarms are generated in the initial rest phase, i.e. $f(t)$ remains below the thresholds during the first 120 s. Note that the relevant information is generated when $f(t)$ crosses for the first time the horizontal line corresponding to the threshold and remains above it for the subsequent 5 s. After this event occurs, the data interpretation procedure attaches no specific information to the behavior of $f(t)$, which, especially after the occlusion has been removed, is strongly influenced by transient phenomena. The results of the tests show that the device correctly monitors the variations of concentration of the oxygenated hemoglobin and of the deoxygenated hemoglobin and that it is possible to choose a suitable threshold to detect the establishment of ischemic conditions.

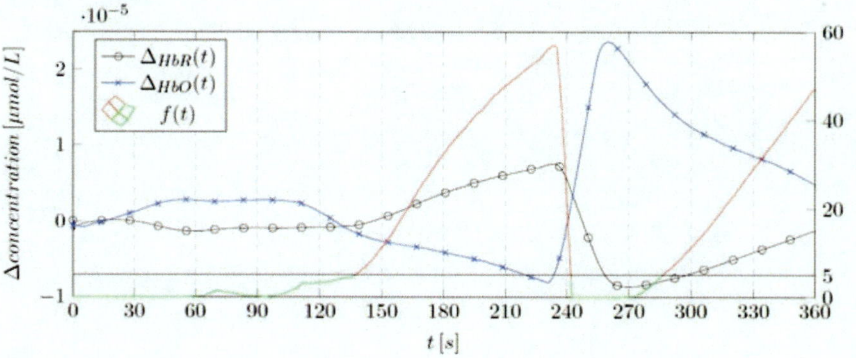

Fig. 3 Results of the test 2 in dry conditions

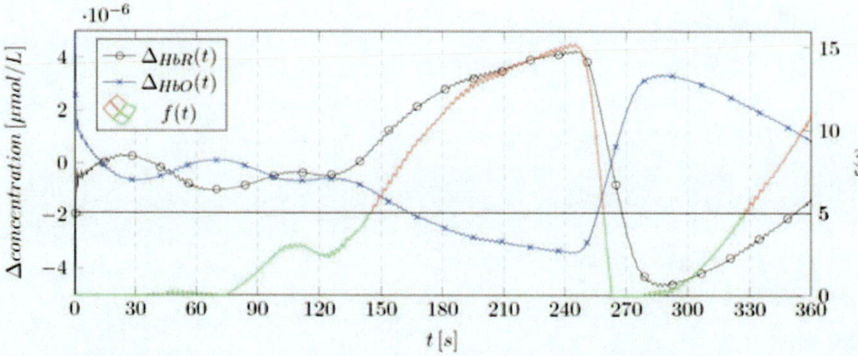

Fig. 4 Results of the test 3 in dry conditions

4.2 Tests in Water

Since the device is conceived to be used in the underwater environment, one of the concerns of this study is to determine if the presence of a layer of water between the probe and the tissue can substantially affect measurements and experimental results. For that purpose, the above test has been repeated in water, simply by keeping the arm of the subject in a water-filled small container. In the first of the two tests reported for sake of illustration, the water temperature was 14 °C, while in the second one, it was 36 °C. Except for the presence of the water, the procedure was identical to the one described in the previous subsection.

The result obtained in the first test is reported in Fig. 5. The first phase of the test is characterized by a large decrease in the concentration of oxygenated hemoglobin, probably due to the physiological response when the arm is immersed in cold water. This behavior makes it difficult to identify the occlusion using the oxygenated hemoglobin signal, but the deoxygenated hemoglobin seems to be less affected by this phenomenon and the data interpretation procedure works properly with a

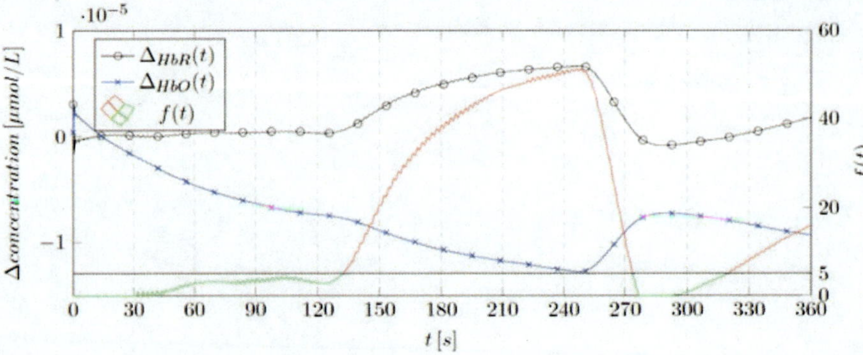

Fig. 5 Results of the test in cold water

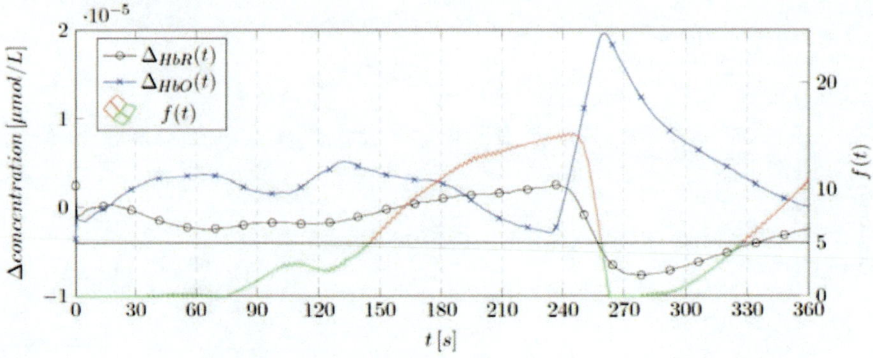

Fig. 6 Results of the test in warm water

29 s(= 12 s + 12 s + 5 s) delay. The value of $f(t)$ was close to exceed the threshold before the occlusion, but the test was globally successful. The result of the second test is reported in Fig. 6. Here, there is no initial decrease of the oxygenated hemoglobin in the initial phase, and, also, in this case, the occlusion is correctly detected by the data interpretation procedure with a total delay of 46 s. The test is successful, although the delay is close to the maximum acceptable one.

5 Conclusions

A portable and inexpensive underwater NIRS device has been designed and constructed and its ability to detect critical ischemic conditions by means of a suitable data interpretation procedure has been tested. The first tests in the laboratory have been successful, and a more extensive experimental campaign for assessing the correct functioning of the device in the field and in various scenarios is planned for the next future. In particular, the performance of the device in monitoring the variations of concentration of the oxygenated and deoxygenated hemoglobin in regions of the brain will be the object of the study.

References

1. Amendola C, Lacerenza M, Buttafava M, Tosi A, Spinelli L, Contini D, Torricelli A (2021) A compact multi-distance DCS and time domain NIRS hybrid system for hemodynamic and metabolic measurements. Sensors 21(3):870
2. Bauernfeind G, Leeb R, Wriessnegger SC, Pfurtscheller G (2008) Development, set-up and first results for a one-channel near-infrared spectroscopy system. Biomed Eng 53(1):36–43
3. Boushel R, Piantadosi CA (2000) Near-infrared spectroscopy for monitoring muscle oxygenation. Acta Physiologica Scandinavica 168(4):615–622
4. Chiarelli AM, Perpetuini D, Filippini C, Cardone D, Merla A (2019) Differential pathlength factor in continuous wave functional near-infrared spectroscopy: Reducing hemoglobin-cross talk in high-density recordings. Neurophotonics 6(3):035005
5. Cialoni D, Pieri M, Balestra C, Marroni A (2017) Dive risk factors, gas bubble formation, and decompression illness in recreational scuba diving: Analysis of DAN Europe DSL data base. Front Psychol 8:1587 (2017). https://www.frontiersin.org/article/10.3389/fpsyg.2017.01587
6. Coyle S, Ward TE, Markham C (2003) Cerebral blood flow changes related to motor imagery, using near-infrared spectroscopy (NIRS). In: World congress on medical physics and biomedical engineering (2003)
7. Coyle SM, Ward TE, Markham CM (2007) Brain-computer interface using a simplified functional near-infrared spectroscopy system. J Neural Eng 4(3):219
8. Delpy DT, Cope M, Van Der Zee P, Arridge S, Wray S, Wyatt J (1988) Estimation of optical pathlength through tissue from direct time of flight measurement. Phys Med Biol 33(12):1433
9. Eichhorn L, Erdfelder F, Kessler F, Doerner J, Thudium MO, Meyer R, Ellerkmann RK (2015) Evaluation of near-infrared spectroscopy under apnea-dependent hypoxia in humans. J Clin Monit Comput 29(6):749–757
10. Gerovasili V, Dimopoulos S, Tzanis G, Anastasiou-Nana M, Nanas S (2010) Utilizing the vascular occlusion technique with NIRS technology. Int J Ind Ergon 40(2):218–222
11. Grassi B, Quaresima V (2016) Near-infrared spectroscopy and skeletal muscle oxidative function in vivo in health and disease: a review from an exercise physiology perspective. J Biomed Opt 21(9):091313
12. Guyton GP, Stanek KS, Schneider RC, Hochachka PW, Hurford WE, Zapol DG, Liggins GC, Zapol WM (1995) Myoglobin saturation in free-diving Weddell Seals. J Appl Physiol 79(4):1148–1155, 091313
13. Jobsis FF (1977) Noninvasive, infrared monitoring of cerebral and myocardial oxygen sufficiency and circulatory parameters. Science 198(4323):1264–1267
14. Jones B, Dat M, Cooper CE (2014) Underwater near-infrared spectroscopy measurements of muscle oxygenation: laboratory validation and preliminary observations in swimmers and triathletes. J Biomed Opt 19(12):127002
15. Jones S, Chiesa ST, Chaturvedi N, Hughes AD (2016) Recent developments in near-infrared spectroscopy (NIRS) for the assessment of local skeletal muscle microvascular function and capacity to utilise oxygen. Artery Res 16:25–33, 127002
16. Kohl M, Nolte C, Heekeren HR, Horst S, Scholz U, Obrig H, Villringer A (1998) Determination of the wavelength dependence of the differential pathlength factor from near-infrared pulse signals. Phys Med Biol 43(6):1771, 127002
17. Lange F, Tachtsidis I (2019) Clinical brain monitoring with time domain NIRS: a review and future perspectives. Appl Sci 9(8):1612, 127002

18. Li T, Zhong F, Pan B, Li Z, Huang C, Deng Z (2017) A brief review of opt101 sensor application in near-infrared spectroscopy instrumentation for intensive care unit clinics. Sensors 17(8):1701
19. Martin DS, Levett DZ, Bezemer R, Montgomery HE, Grocott MPW (2013) the Caudwell Xtreme Everest Research Group: the use of skeletal muscle near infrared spectroscopy and a vascular occlusion test at high altitude. High Alt Med Biol 14(3):256–262, 127002
20. McKnight JC, Bennett KA, Bronkhorst M, Russell DJ, Balfour S, Milne R, Bivins M, Moss SE, Colier W, Hall AJ et al (2019) Shining new light on mammalian diving physiology using wearable near-infrared spectroscopy. PLoS Biol 17(6):e3000306
21. Naseer N, Hong KS (2015) fNIRS-based brain-computer interfaces: a review. Front Hum Neurosci 9:3, e3000306
22. Petrantonakis PC, Kompatsiaris I (2018) Single-trial NIRS data classification for brain-computer interfaces using graph signal processing. IEEE Trans Neural Syst Rehabil Eng 26(9):1700–1709, e3000306
23. Roldán M, Kyriacou PA (2021) Near-infrared spectroscopy (NIRS) in traumatic brain injury (TBI). Sensors 21(5):1586
24. Scheeren TWL, Schober P, Schwarte LA (2012) Monitoring tissue oxygenation by near infrared spectroscopy (NIRS): background and current applications. J Clin Monit Comput 26(4):279–287, e3000306
25. Submersible NIRS: a very wet lab. https://www.artinis.com/blogpost-all/2019/submersible-nirs-a-very-wet-lab. Accessed 13 Feb 2021
26. Si J, Zhang X, Li M, Yu J, Zhang Z, He Q, Chen S, Zhu L, Jiang T (2021) Wearable wireless real-time cerebral oximeter for measuring regional cerebral oxygen saturation. Sci China Inf Sci 64(1):1–10, e3000306
27. Suzuki S, Takasaki S, Ozaki T, Kobayashi Y (1999) Tissue oxygenation monitor using NIR spatially resolved spectroscopy. In: Optical tomography and spectroscopy of tissue III. vol 3597, pp 582–592. International Society for Optics and Photonics, SPIE. https://doi.org/10.1117/12.356862
28. Wang D, Wang L, Wang J, Gao P, Li S, Zhu L, Li Z, Chen T, Li F, Feng H (2021) Characterization of the ischemic muscle by quantitative hybrid diffuse optical measurement. Opt Commun 483:126579
29. Williams CL, Meir JU, Ponganis PJ (2011) What triggers the aerobic dive limit? Patterns of muscle oxygen depletion during dives of emperor penguins. J Exp Biol 214(11):1802–1812, 126579
30. Yu G, Durduran T, Lech G, Zhou C, Chance B, Mohler ER, Yodh AG (2005) Time-dependent blood flow and oxygenation in human skeletal muscles measured with noninvasive near-infrared diffuse optical spectroscopies. J Biomed Opt 10(2):024027

Pill Defect Detection Using an Improved Convolutional Neural Network

Thi Thoa Mac

Abstract A novel effective method to detect pill defects during pill manufacturing is proposed in this study. We have developed an analysis program that incorporates deep learning convolutional neural networks to fully automate the image analysis of pills for internal crack detection. The deep learning tool based on YOLO algorithm is effectively implemented into the industrial pharmaceutical workflow. Firstly, we analyze Gauss filtering and smoothing techniques for pill detection. Secondly, Hog feature extraction is introduced to simplify the representation of the image that contains only the most important information about the image. Lastly, improved YOLO is designed for online detection of pill defects. The proposed approach obtains robust quantification of internal pill cracks.

Keywords Quality Inspection · YOLOv3 · Deep learning

1 Introduction

The demand for real-time defect detection on the high-precision manufacturing system has increased with the intelligent technique in industrial 4.0. Considering that pharmaceutical pills manufacture has a wide market. The high accuracy in pills defect detection production line is an important factor to advance the level of the manufacturing [1].

The traditional approaches are non-computer vision-based approaches, such as the 'Pillbox' by the United States National Library of Medicine [2], 'Pill Identifier' by Medscape [3], and 'Pill Identification tool' by WebMD [4]. The manual input of these systems are shape, color, imprints, and scoring. The disadvantages of these approaches are time consuming, laborious, and subjective.

Besides, pills defect detection based on machine vision is an advanced method in which the contour extraction algorithm is implemented to extract the image

T. T. Mac (✉)
School of Mechanical Engineering, Hanoi University of Science and Technology, No. 1, Dai Co Viet, Hai Ba Trung, Hanoi, Vietnam
e-mail: thoa.macthi@hust.edu.vn

G. Conte and O. Sename (eds.), *Proceedings of 10th International Conference on Mechatronics and Control Engineering*, Lecture Notes in Mechanical Engineering,
https://doi.org/10.1007/978-981-19-1540-6_7

feature. The next stage is a template matching process [5, 6]. This approach has low speed as well as low efficiency as they heavily depend on the lighting condition. As consequence, the accuracy is dramatically decreased.

Deep learning is a powerful tool in various real life and industrial applications. To archive the high quality of real-time defect detection, the essential requirements are fast detection processing and high classification accuracy. The most popular deep learning approaches are You Only Look Once (YOLOv3) [7], FAST-RCNN [8, 9], SSD, Deep Convolutional Network (DCN) [10], and FPN [11]. In which, YOLOv3 uses an end-to-end method to regress features to reduce computational burden. Several studies indicate that YOLOv3 is more advantage than other approaches.

In this study, we investigate a improved CNNs to fully automate detect internal crack/contamination pills in the manufacturing system. Firstly, Canny and Gauss filterings are developed for pill detection. Secondly, Hog feature extraction is investigated to simplify representation of the image that contains only the most important information. Lastly, improved YOLO is proposed for online detection of pill defects process. The proposed approach produces reference for the pills manufacturing industry to enhance production efficiency, product quality, and reduce laborious.

This paper is organized as follows. In Sect. 2, we explain the system design in Sect. 2.1. In addition, the improved CNNs structure is introduced in Sect. 2.2. In Sect. 3, we give a description of experiment results and discussions. Finally, conclusion and future work are provided in Sect. 4.

2 Setup System and Methodology

2.1 The System Design

The system is designed in lab-scale as shown in Fig. 1 (Left). The system flowchart of the online platform for pill manufacturing defects is designed by our research group as shown in Fig. 1 (Right). The pills defect detection systems uses camera to take driven pills' images on the belt conveyor. The platform includes the data processor, data acquisition sensor, light source. The workstation of Intel core i7-66600U CPU is used for data analysis. Based on the results, the programmable logic controller (PLC) sends the control signals to the real system. If the pills are defective, they are excluded from the system. Otherwise, the qualified pills are pushed into the bottle at a packaged module.

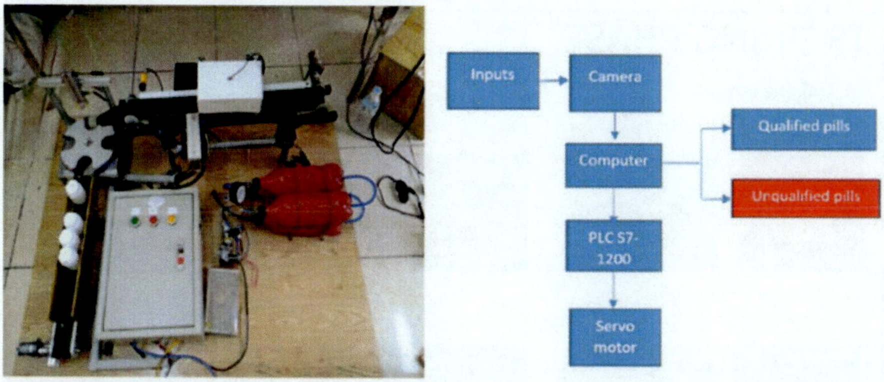

Fig. 1 **Left**. The pills defect detection system design; **Right**. Flowchart of the experimental platform

2.2 *Improved Construction and Training of YOLOv3 Network*

The improved YOLO structure is presented in Fig. 3. Combining the types and characteristics of defects during actual pills manufacturing and the complex background of image acquisition on the factory production line, an improved online defect detection network for YOLOv3 pill is proposed. In this network structure, the end-to-end Darknet-53 convolutional network formed in YOLOv3 is maintained. In the smoothing layer, an Canny filter is used for edge detection. Then, a Gaussian filter with a pixel of 3×3 is then used for the secondary smoothing of the image. This filter mainly reduces the high-frequency noise in the collected image and further reduces the dust particles, and other impurities in the pills production system (Fig. 2).

The pills inspection using YOLO algorithm has two phases: training phase and validation phase. Training/validation phase includes four steps: Training data input (raw image), Canny for edge detection, Gauss Filter for Hog Feature Extraction,

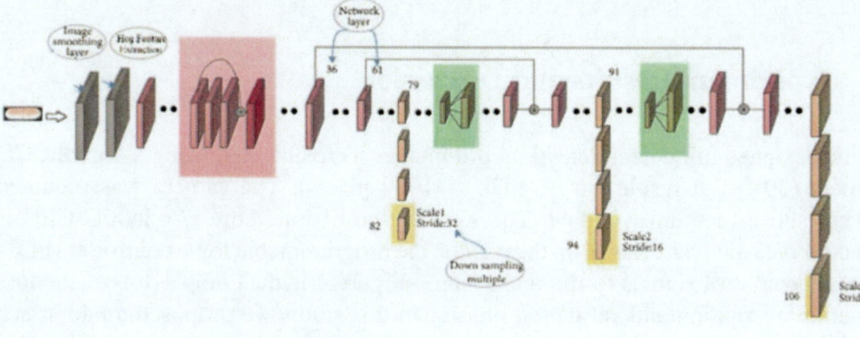

Fig. 2 Modified YOLO detection structure

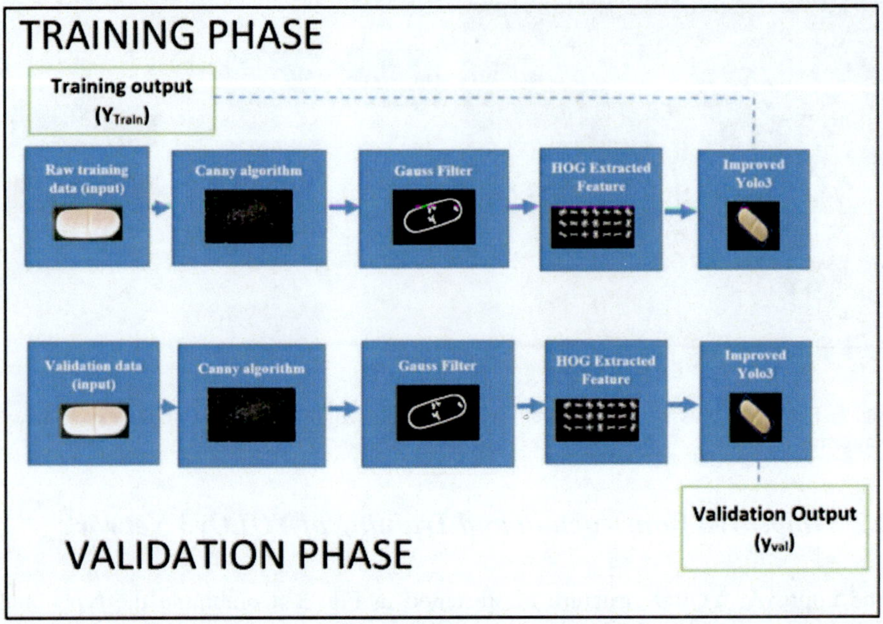

Fig. 3 Pill inspection using a machine learning algorithm has two phases: training phase and test phase

and Classification/Regression/Clustering. The feature extraction uses the feature descriptor from Hog algorithm. In our application, the training architecture includes:

- Number of epochs: 110
- Batch size: 8
- Train data/Validation data: 4:1
- Optimizer: Adam with learning rate = 0.5 × 10 − 4 in the first 20 epochs, learning rate = 0.5 × 10 − 5 in the rest epochs. The purpose is to quickly convert to the optimal value.

3 Experiment Results and Discussion

In the designed pill defect detection, pill images were obtained using a digital CCD camera (30 fps at resolution of 1920 × 1080 pixels). The camera was mounted 20 cm above the conveyor line. The workstation of Intel core i7-66600U CPU is used for data analysis. Based on the results, the programmable logic controller (PLC) sends the control signals to the real system. Figure 4 is the comparison of the loss function of training and validation phases. In this study, we proposed an approach that integrates YOLOv3, image processing to fully automate and detect the internal crack/contamination/good quality pills. In which, we first analyzed Canny and Gauss

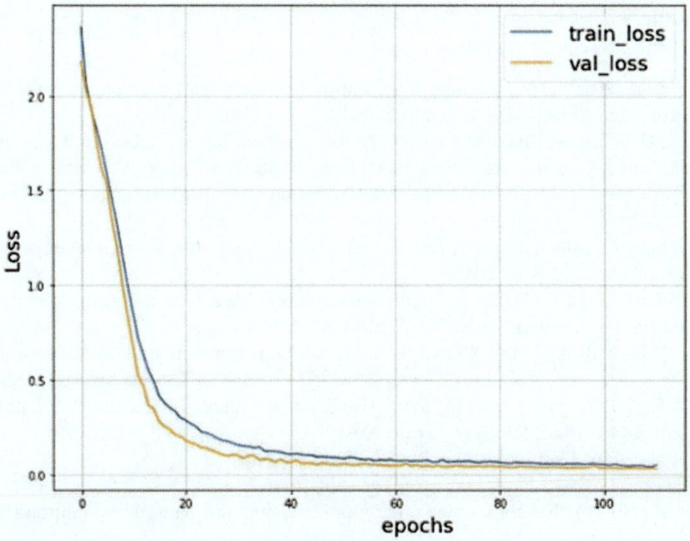

Fig. 4 The comparison of training and validation loss

filters for pill detection. Secondly, Hog feature extraction is introduced to simplify representation of the image that contains only the most important information about the image. Lastly, improved YOLO v3 is designed for online detection of pill defects.

4 Conclusion

In this paper, we have developed a pill defect detection in a lab-scale system to validate the proposed improved convolutional neural network. In addition, an analysis program that incorporates deep learning convolutional neural networks to fully automate the image analysis of pills for internal crack detection is investigated. The deep learning tool based on image processing is effectively implemented into the designed workflow system. Firstly, we first analyzed Canny algorithm and Gauss filtering for pill detection. Secondly, Hog feature extraction is implemented to simplify representation of the image. Lastly, improved YOLO v3 is proposed for online detection of pill defects. The proposed approach obtains robust quantification of internal pill cracks with proper loss function values.

Regarding future work, the proposed approach will be investigated in a large-scale pill manufacturing inspection, which brings the advantages of the production time and production cost reduction.

Acknowledgements This work was supported by the Autonomous Higher Education Project (SAHEP) grant funded under number T2020-SAHEP-012.

References

1. Tobore I et al (2019) Deep learning intervention for health care challenges: some biomedical domain considerations. JMIR Mhealth Uhealth 7(8):e11966 (2019)
2. Pillbox (2021). http://pillbox.nlm.nih.gov/pillimage/search.php. Accessed 9 Mar 2021
3. Pill Identifier (2021). http://reference.medscape.com/pill-identifier. Accessed 9 Mar 2021
4. Pill Identification Tool (2021). http://www.webmd.com/pill-identification. Accessed 9 Mar 2021
5. Guo Q, Zhang C, Liu H, Zhang X (2016) Defect detection in tire X-ray images using weighted texture dissimilarity. J Sens 2016
6. Yuan J, Wang Q, Li B (2014) A flexile and high precision calibration method for binocular structured light scanning system. Sci World J 2014(8)
7. Redmon J, Farhadi A (2018) Yolov3: an incremental improvement. http://arxiv.org/abs/1804.02767
8. Girshick R (2015) Fast r-cnn. In: 2015 IEEE international conference on computer vision (ICCV), pp 1440–1448, Santiago, Chile, Dec 2015
9. Liu W et al (2016) Lect Notes Comput Sci 9905:21
10. Sabri AH et al (2018) J Drug Deliv Sci Technol 46:16
11. Lin T et al (2017) IEEE conference on computer vision and pattern recognition (CVPR), vol 2117

Application of Method Engineering Tools to Improve the Productivity of the Production System in the Textil Andes Company

A. F. Grimaldo, J. A. Machacuay, and E. D. Vilchez

Abstract This paper on pre-experimental design proposes the application of method engineering tools, focused on improving the productivity of the Textil Andes company dedicated to the manufacture of ethnic ribbons. By identifying deficiencies and bottlenecks, through the use of the operation process chart and routing diagram, the aim was to reduce the distances and times used; in fact, the processes are unique in the region, therefore it was decided to calculate their standard time which was 250.25 min, and that by implementing a new distribution and working method it managed to reduce the standard time by 11.65% and the distances used by the product process flow by 49.51%; consequently, a less production time per batch added to the management of a periodic preventive maintenance format, generating a significant decrease in defective units; in other words, the number of good units produced, measured by efficacy, increased by 11.59%. Likewise, the reduction in idle time caused an increase from 89.07 to 94.28% in efficiency, thus achieving an improvement in productivity of 14.77% that directly influences attention and production capacity.

Keywords Bottleneck · Efficacy · Efficiency · Method engineering · Productivity

1 Introduction

The global shock of SARS-CoV-2 caused great repercussions in the economic sector, especially in the textile industry [1], in which 10% of the Gross Domestic Product

A. F. Grimaldo (✉) · J. A. Machacuay
Universidad Continental, Huancayo 12001, Peru
e-mail: 74299141@continental.edu.pe

J. A. Machacuay
e-mail: 70034217@continental.edu.pe

E. D. Vilchez
Universidad Federal del Norte (Artico) M.V. Lomonosov, 198411 Lomonosov, Russia
e-mail: evilchez@continental.edu.pe

G. Conte and O. Sename (eds.), *Proceedings of 10th International Conference on Mechatronics and Control Engineering*, Lecture Notes in Mechanical Engineering,
https://doi.org/10.1007/978-981-19-1540-6_8

(GDP) of the manufacturing sector in the country is included in the industry of textiles and clothing [2], where Peruvian companies strive to survive a critical recession in exports that indicate the 25 million dollars that took place in 2020, compared to 120 million dollars in 2019 [3]. In addition, the SMEs are the most affected due to the situation they face as they continue to survive in a market that forces them to make significant cuts in personnel and productive capacity, consequently, their productivity, which led to the failure of many of them [4]. On the other hand, the government poses the challenge of reactivating the textile sector by calling for the purchase of Peruvian products internationally; however, it is pointed out that we are perfectly supplied for local demand but not for global [5]; for that reason, companies and micro-enterprises, even with the situation against them, must begin to propose strategies to improve their productivity level even with fewer resources [6], so that the reactivation plan is beneficial in economic terms. Therefore, the method engineering application model performs a corrective analysis, through a data record, indicators and operational diagrams [7], which greatly favors these small manufacturing companies in a shorter time, focusing directly on all the factors that negatively influence production and implementation of immediate changes in the processes in order to optimize them and adapt them to continuous improvement [8, 9].

2 Methodology

The objective of method engineering is to reduce unproductive times and movements, in order to increase productivity in any manufacturing plant [10]. To identify the deficiencies in the process, which significantly affected the productivity of the company [11], the causes can be observed using the Ishikawa diagram (see Fig. 1).

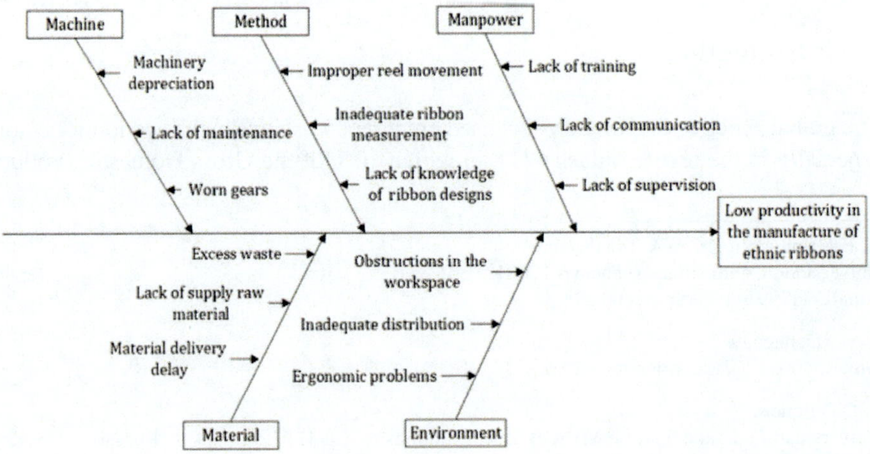

Fig. 1 Ishikawa diagram

2.1 Select the Product Process to Be Studied

The population was taken as all the processes for the manufacture of ethnic ribbons and as a sample of the processes involved in the product with the highest demand [12], which was identified with the help of the bar chart, based on historical data from the first trimester of 2020.

2.2 Collect Information

Data collection was done by direct observation and the use of method engineering tools: (a) Operation process chart shows the chronological sequence of all operations and component input, from raw material to final product packaging. (b) Routing diagram provides a graphical and sequential view of the flow of all operations and distances on a metric scale. (c) Standard time format, where the times observed in a given period were recorded, then the average time was obtained, which is multiplied by the valuation factor (performance rating) and supplements (time to carry out normal and physiological needs) to determine the standard time of the production process [13].

$$Standardtime = Observedtime \times Valuationfactor \times (1 + supplements) \tag{1}$$

2.3 Critical Examination

It is the verification of the information obtained in relation to the problem presented, through an objective analysis that recognizes the productive capacity of the company [14]. The following formats are used.

Efficiency format. Idle times are identified for the calculation of the useful time, which is divided by the programmed time to obtain the efficiency rate [13].

$$Efficiency = \frac{UsefulTime}{ProgrammedTime} \tag{2}$$

Efficacy format. The units produced minus the defective units in a period are recorded, then divided by the programmed units to find the efficacy rate. [13].

$$Efficacy = \frac{Unitsproduced - Defectiveunits}{Programmedunits} \tag{3}$$

Productivity format. The indicators of distance traveled, standard time, units produced, efficiency and efficacy were considered. Therefore, the determination of productivity is reflected in a more specific and clearer way [13].

$$Productivity = Efficacy \times Efficiency \tag{4}$$

2.4 Devising and Defining Methods

It consists of determining the new working methods in each cycle, looking for ways to include innovative elements, improved aspects and other perspectives that eliminate bottlenecks [15, 16].

2.5 Implementation of the Methods

It is ensured that the new methods can be implemented, in other words, the organization and execution of all the activities involved in the proposal are carried out by using the schedule of activities according to the availability of the company for the development of the methods [17], to avoid interruptions and inconveniences in the development of their usual tasks.

2.6 Evaluation of the New Methods

The results and effects of the new methods are monitored, through the comparison of pre-test and post-test results of the established indicators [18].

3 Results

3.1 Selection of the Most Demanded Product

A study was done on the demand for all the products offered by the company, which are grouped according to their dimension: 1.10, 1.50, 2.50, 3.50 and 8 cm. After data collection, Fig. 2 shows that the 1.5 cm product is the one with the highest demand.

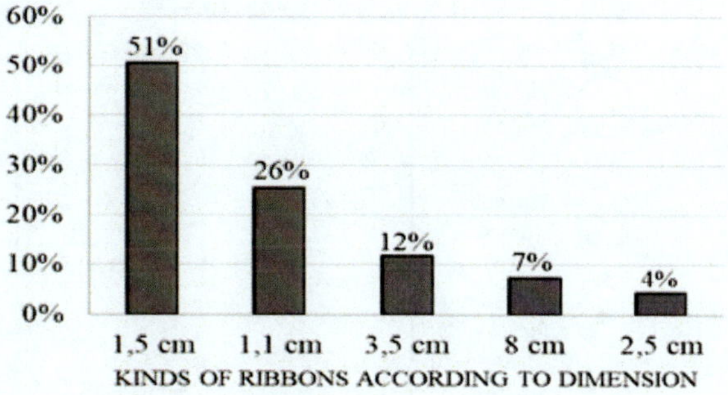

Fig. 2 Demand for the products of the Textil Andes company

3.2 *Compilation of Times and Movements*

Manufacturing process cycles. The manufacturing process of ethnic ribbons begins in the preparation cycle where the thread cones are selected and put according to the ribbon design; they pass through a series of holes until they reach the empty reel, which is placed on the press of the winding machine. Subsequently, the winding cycle is performed when the threads are distributed on a reel and programmed according to the number of meters requested to be winded, where the operator manages the speed and monitors if any thread is finished or on breakage of this, once the reel is filled, the threads must be adjusted and carried to weaving area, for which the full reel is placed into the weaving machine and the threads are put in the appropriate holes. Finally, the measurement cycle, used the meters of ribbons are manually measured in 11 m, which will be joined in groups of 10 to be packed in bags. The following is the operation process chart (see Fig. 3), which shows the sequence of activities in the 1.5 cm ribbon manufacturing system.

Register standard time. Data were taken from the execution times for each cycle, with a valuation of 95% and supplements of 14%; applying Eq. (1), 250.25 min was obtained as the standard time (Table 1).

Layout plant. By showing the layout of the production plant (see Fig. 4), it is possible to visualize the route required for the flow of each ribbon production cycle.

3.3 *Critical Analysis of Production Indicators*

Efficacy. It is programmed to produce 6000 m of ribbons per day, however, on average 4579.45 m were produced, of which 161.90 were defective; applying Eq. (2), we obtain an efficacy rate of 73.63%.

Efficiency. There are 480 min available per day, but in the 15 days of the study, 52.48 min of idle time were obtained, so the useful time was 427.52 min. Applying Eq. (3), 89.07% efficiency was obtained.

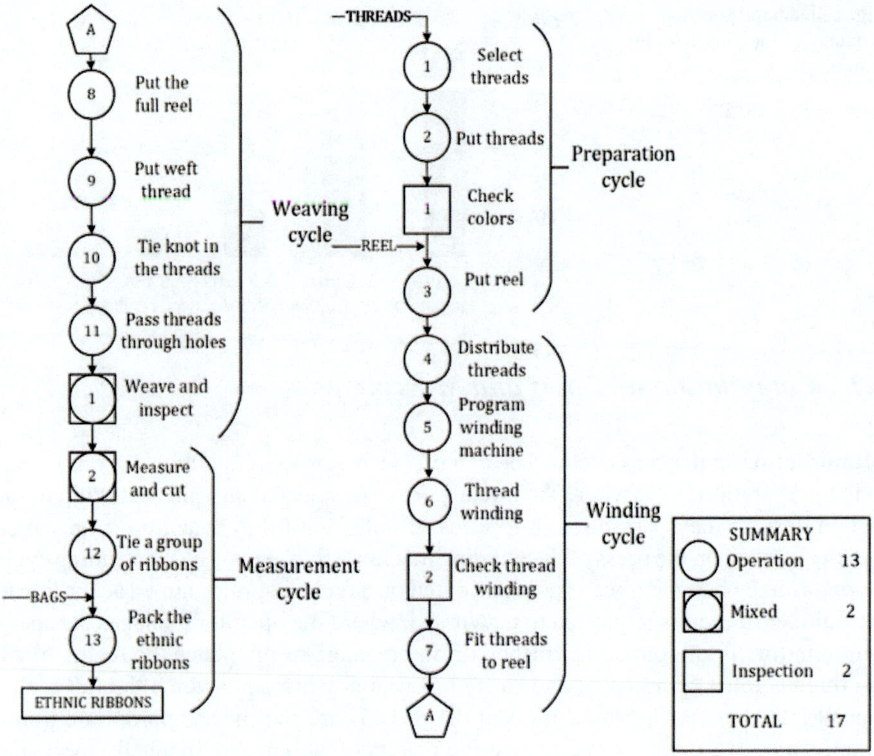

Fig. 3 Operation process chart of the ethnic ribbon manufacturing

Table 1 Standard production process time

Times	Production cycles				Total (min)
	Preparation cycle	Winding cycle	Weaving cycle	Measurement cycle	
Average time	33.895	86.712	81.382	29.081	231.07
Assessment	0.950	0.950	0.950	0.950	
Normal time	32.200	82.376	77.313	27.627	219.52
Supplement	0.14	0.14	0.14	0.14	
Standard time	36.708	93.909	88.137	31.494	250.25

Productivity. The productivity indicator involves efficiency and efficacy. With the above data and the application of Eq. (4), 65.58% productivity was obtained.

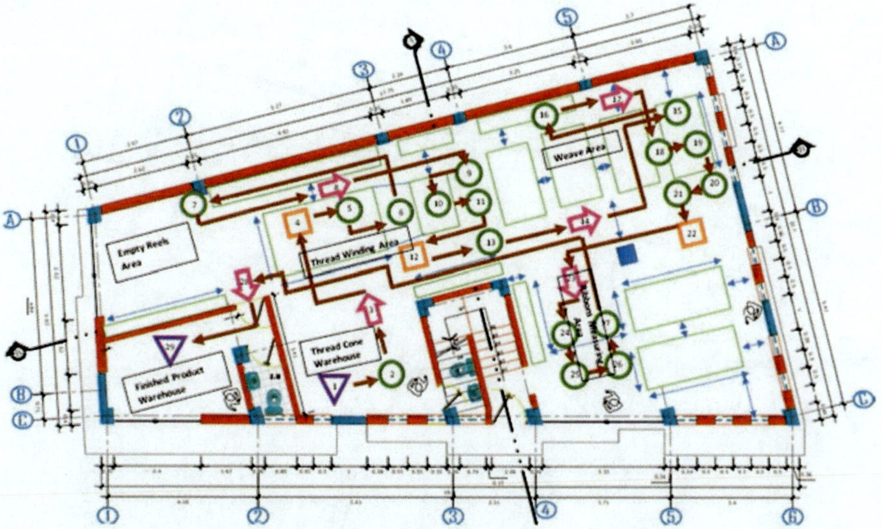

Fig. 4 Routing diagram of the ribbon production system

3.4 *Approach to the New Methods and Distribution*

Application of the new methods. The following methods were applied for each cycle: (a) Preparation cycle, the cones of threads are placed on the pendants according to the ribbon design; the color palette was implemented with the necessary information to speed up the process and avoid confusion. (b) Winding cycle, at the moment of winding the threads, they break or end, so a signaling of the route was made for quick attention of any error. At the end of the winding process, a conveyor cart was implemented to take the full reel of 20 kg to the weaving area, to avoid possible injuries. (c) Weaving cycle, it should be noted that the machines need attention to avoid idle time, so a preventive maintenance format was implemented to control the functions of the weaving and winding machine, since stopping production contributes to idle time, which we propose to mitigate. (d) Measurement cycle, there is no specific space for measuring ribbons; therefore, a suitable space was located and equipped with the necessary materials for its execution.

New distribution. For the new distribution, as it is a small company, a criterion based on product flow and factors that influence the route for its production was used, maintaining the sequentially of the processes and oriented in a single direction, which obtains clearer areas of displacement and shorter route distances, as can be seen in Fig. 5.

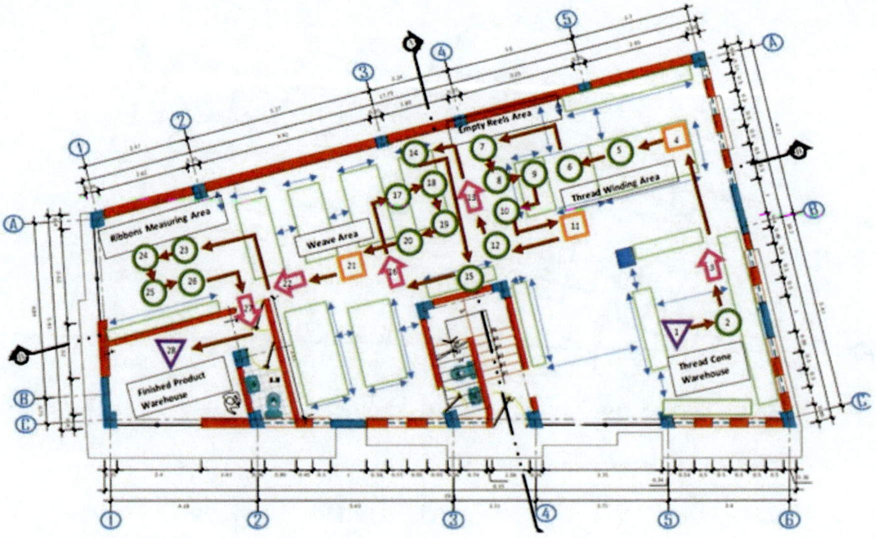

Fig. 5 Routing diagram with the new distribution

Table 2 Implementation schedule

Activities	Jul	Aug	Sept	Oct	Nov	Dec	Jan
1. Distribution approach	X						
2. Planning of new methods	X						
3. Communication of the implementation to personnel		X					
4. Resource preparation and conditioning			X	X			
5. Implementation of the new distribution				X			
6. Staff training					X		
7. Implementation of new methods						X	
8. Improvement review					X	X	X

3.5 Implementation of the New Distribution and Methods

For the implementation, the activities were scheduled according to the availability of the company, from July 15, 2020, to January 26, 2021 (Table 2).

3.6 Implementation Evaluation

By applying the new methods, the standard time was reduced from 250.25 to 221.10 min. In addition, by implementing the new distribution, a better route flow

Table 3 Production indicators

Production indicators	Results before	Results after
Useful time	427.52	452.52
Total time	480	480
Quantity produced	4579.45	5168.34
Quantity programmed	6000	6000
Efficiency	0.8907	0.9428
Efficacy	0.7363	0.8522
Productivity	0.6558	0.8035

was obtained, and the distance traveled by the product decreased from 192.60 m to 97.24 m. Table 3 shows that the production indicators improved, efficiency increased from 89.07 to 94.28% and efficacy from 73.63 to 85.22%, which gives an increase in productivity from 65.58 to 80.35% and shows an improvement after the application of the new layout.

4 Conclusions and Recommendations

According to the study applied to the company, it was concluded, through the historical data of the first trimester of 2020, that the most demanded product is the 1.5 cm ribbon, whose production system had a standard time of 250.25 min and a distance traveled in the production process of 192.60 m, which were recorded by means of the operations process chart and the route diagram focused on the bottlenecks in each production cycle giving a faster and more complete result of the deficiencies of the company. Subsequently, with the implementation of the new distribution and work methods, larger workspaces were obtained for each cycle, better flow of movement and optimization of time in the execution of tasks, thus reducing the standard time by 11.65% and distances routed by 49.51%. On the other hand, the production indicators improved after the implementation, resulting in a reduction of idle times, which helped to increase efficiency from 89.07 to 94.28%, also the number of defective units decreased, which contributed to increased efficacy from 73.63 to 85.22%, as a result of which productivity increased by 14.77%. Therefore, it can be concluded that the application of method engineering tools significantly improves the production system of the ethnic ribbon of the Textil Andes company. Due to the relevance of the data collected, it gives rise to continue improving the research to involve other factors such as indicators of machine availability and performance, which provide a broader picture of the scope of the application of method engineering and its integral use of the same, in addition, a study where the quality factor is complemented, due to the fact that when applying method engineering it is required to improve times and speed up operations, so it is frequent that the quality of the products is affected.

References

1. International Labour Organization 2020 COVID-19 and the textiles, clothing, leather and footwear industries. International Labour Organization, Geneva, pp 1–7
2. Posada C (2020) Textile sector must take advantage the TLC to win market in the USA. La cámara, Lima p 16
3. Reaño M (2020) The challenge of reactivating the textile industry. Industrial eEngineering. Universidad de Lima, Lima, pp 1–2
4. Instituto de Economía y Desarrollo Empresarial (2020) Peruvian economy would fall 3.7 % this year by effect of covid-19. Instituto de Economía y Desarrollo Empresarial. La cámara, Lima, pp 14–16
5. Gestión D (2020) Internal and external scenario of the textile industry. Business office. Diario Gestión, Lima
6. Kaur K (2021) The early impact of COVID-19 on textile industry: an empirical analysis. Rev Manag Labor Stud 2–13
7. Vildes E, Díaz L, Gutiérrez J (2017) Methodological analysis for the performance of studies of methods and times. Res Dev TIC 8:3–10
8. Andrade Y, Cardenas L, Viacava G (2019) Lean manufacturing model for the reduction of production times and reduction of the returns of defective items in the textile industry. In: International conference on applied human factors and ergonomics, pp 387–398
9. Garza J, Kumar V, Chaikittisilp S, Hua Tan K (2018) The effect of lean methods and tools on the environmental performance of manufacturing organizations. Int J Prod Econ 200:170–180
10. Andrade A, Del Río C, Alvear D (2019) A study on time and motion to increase the efficiency of a shoe manufacturing company. Información tecnológica 30:83–94
11. Ramírez Y, Castellares R (2018) Study of times and movements to improve the productivity of a fishing company. INGnosis J Sci Res 4:64–77
12. Grimaldo A, Machacuay J (2020) Proposal to improve the production system of ethnic ribbons based on the application of method engineering tools to increase productivity in the Textil Andes Company. Universidad Continental, Huancayo, pp 1–94
13. Garcia R (2005) Work study: methods engineering and work measurement, 2nd edn. McGraw Hill, Mexico
14. Bobir T (2017) Features of the method of calculation of production capacities of the textile enterprises. Бюллетень науки и практики 10:213–222
15. Montaño K, Preciado J, Robles J, Chávez L (2018) Methods of work to improve the competitiveness of the Sonora's table grape system. Alimentación Contemporánea y Desarrollo Regional 28:2–25
16. Valdivieso B, Meza H, Gutiérrez E (2019) Application of the improvement of working methods to increase the productivity of the anchovy fillet production process. INGnosis J Sci Res 5:113–125
17. Thomas T, Sherman SR, Sawhney RS (2018) Application of lean manufacturing principles to improve a conceptual 238Pu supply process. J Manuf Syst 46:1–12
18. Cristobal J, Quispe G, Dominguez F (2020) Waste reduction with Lean manufacturing model in an alpaca wool workshop. Int J Prod Econ 999:1–7

Design and Experimental Validation of an H_∞ Adaptive Cruise Control for a Scaled Car

Wissam Sayssouk, Hussam Atoui, Ariel Medero, and Olivier Sename

Abstract Adaptive Cruise Control algorithms have been developed as a consequence of recent advances in automobile technology, such as sensors and on-board processing, which increase both comfort and safety. Through this paper, a longitudinal dynamic model is identified based on a collected experimental data. Moreover, the H_∞ concept is used to design a controller to regulate the longitudinal behavior of an autonomous scaled car while ensuring the safety distance between vehicles to avoid collision. The H_∞ control design is based on solving an optimization problem with guaranteed performances. The proposed approach is then implemented and experimentally validated on a real scaled car.

Keywords Adaptive Cruise Control · Autonomous Vehicles · H_∞ control

1 Introduction

In the last decade, self-driving cars have become a popular topic of research. Intelligent and autonomous vehicles are at the heart of the societal concerns of future transportation. These cars, equipped with numerous sensors and actuators, will enhance road safety, streamline traffic, make transportation more accessible to people with disabilities, and participate in the development of new modes of transportation. Multi-objective speed control of automated vehicles has already been studied

This work has been partially supported by the LabEx PERSYVAL-Lab (ANR-11-LABX-0025-01) funded by the French program Investissement d'avenir

W. Sayssouk · H. Atoui · A. Medero · O. Sename (✉)
University Grenoble Alpes, CNRS, Grenoble INP, GIPSA-Lab 38000 Grenoble, France
e-mail: olivier.sename@grenoble-inp.fr

H. Atoui
Research Department, Renault SAS, 1 Avenue de Golf, 78280 Guyancourt, France

A. Medero
Institut de Robòtica i Informàtica Industrial, C/. Llorens i Artigas 4-6, 08028 Barcelona, Spain

G. Conte and O. Sename (eds.), *Proceedings of 10th International Conference on Mechatronics and Control Engineering*, Lecture Notes in Mechanical Engineering,
https://doi.org/10.1007/978-981-19-1540-6_9

considering energy efficiency, fuel consumption, and traveling time. One of the main studies is the design of Adaptive Cruise Control (ACC) strategies for automated car-following. Recent results have proposed new spacing policies and new control architecture to decrease the time gap between vehicles, while ensuring the string stability [1, 2]. Concerning the cruise control performances, the speed control algorithm [3, 4] has been augmented to enhance road stability and safety of the vehicle, see [5]. Moreover, the impact of the look-ahead cruise control on the traffic flow has also been analyzed and a parameter-dependent model has been presented to consider traffic flow in the velocity trajectory design, see [4]. Hence, the design of a longitudinal control strategy becomes challenging.

The objective of the Cruise Control is to track the desired speed provided by the driver. ACC systems are an extension of conventional cruise control (CCC) systems that adjust vehicle velocity and provide a specified distance to the preceding vehicle by automatically controlling the throttle and/or the brake [6]. They rely on numerous number of sensors, i.e., LiDAR, Camera, Radar, etc. Their role is not only to control vehicle speed but also to maintain a safety distance with the preceding vehicle, which is known as a gap distance, while ensuring the string stability in car platooning. ACC with the Stop&Go system offered by the Advanced Driver Assistance System ($ADAS$), makes driving easier and less stressful. Several control approaches in the topic of Adaptive Cruise Control systems have been proposed. The Model Predictive Control (MPC) is largely and commonly used, as in [7] where a high-level controller is proposed based on MPC approaches, taking into consideration the driver longitudinal ride comfort, driver permissible tracking range, and rear-end safety. Carlos et al, [8] developed a Fractional-Order Control (FOC) approach to design a feedforward structure for ACC to enhance the car following while ensuring the robustness/stability. A fuzzy longitudinal control is studied in [10] to control the throttle and the braking pedal, this control approach is using (IF...THEN) conditions based on experience and experimental results. In [9], the Cooperative Adaptive Cruise Control ($CACC$) is considered, it is the extend of the ACC by enabling the wireless communication between vehicles ($V2V$) to control the time gap. An H_∞ Linear Parameter Varying (LPV) approach is used to deal with the variations of the time headway while ensuring the convergence of the spacing errors towards zero and the attenuation of any disturbance propagating along the platoon.

This work considers the H_∞ concept to achieve stabilization with guaranteed performance. First, it is used to design the CC aiming to track the desired speed. Then an ACC H_∞ control strategy is synthesized aiming to follow the preceding vehicle by achieving the desired relative distance with the smallest time headway and standstill distance. Then, both controllers are simulated and experimentally validated on a scaled car.

This paper is organized as follows: Sect. 2 a detailed overview of the studied platform, the electronic components and the car body, Sect. 3 describes the longitudinal vehicle dynamics of the scaled car. The control structures and the H_∞ controllers are designed in Sect. 4 for the Cruise Control and Adaptive Cruise Control, respectively, in addition to the spacing policy. Sections 5 and 6 are devoted for simulation and experimental results, respectively. The conclusions with some suggestions and future works are presented in Sect. 7.

2 Platform Design and Hardware Parts

This section introduces a detailed overview of the studied platform concerning its hardwares and softwares. The vehicle body consists of DFRobot GPX RWD ROB0165 Smart Robot Brushless Motor Racing Car, it is a racing platform mounted with two brushless motors, a front servomotor and a spring suspension allowing a more stable driving. Furthermore, the car has bumpers on both sides, the front and the back to protect it from damage as shown in Fig. 1 (Table 1).

The main components are briefly described below:

Pixy2 Camera: The Pixy2 camera allows the detection and reading of the "road signs" by the barcode detection capabilities using the line tracking mode, which is used later for velocity reference generation.

External Camera and Track: The platform RobotMe at Gipsa-LAB is equipped with an 8 external motion capture camera systems. Those cameras, by Qualysis, have a 3D tracking for each marker (see Fig. 1) attached to the car offering localization for all 6 degrees of freedom (3 for position, 3 for rotation) at a rate of 100 Hz. This system allows measuring the car's position and velocity at each

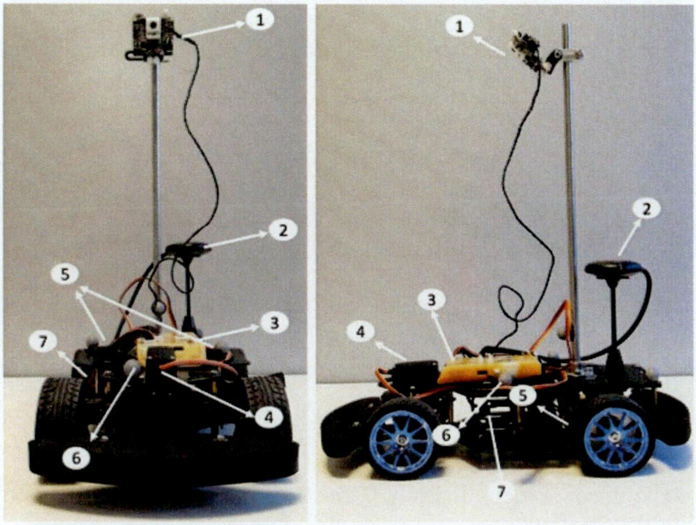

Fig. 1 Front and side view of car components

Table 1 Represents the hardwares equipped to the platform

#	Part	Function
1	Pixy 2 CMUcam5	Barcodes detection
2	GPS HolyBro Pixhawk 4	Turn on/off the rover
3	NXP RDDRONE-FMUK66	Process the data
4	MG996R Servo	For steering
5	2122/13T BLDC	Throttle
6	8 mm Qualisys markers	To track the car position
7	ACCU NI-MH 3000 mAh	Power supply

instant, although its data is not available in real time, so it is used for the model identification and to check the feasibility of the implemented ACC.

Processor Framework: For this project, a PX4 Robotic Drone FMU (fmuk66) is used to control the autonomous car. This board is designed for applications that need high memory densities with low power consumption. It guides and controls the vehicle's navigation as well as its real-time reaction to its surroundings and track. In addition to the feature that comes with the PX4, which allows to log on SD card, a new topic is created to extract the needed data that will be used in the next section to identify the car's longitudinal dynamic model.

3 Longitudinal Vehicle Dynamics

This section presents the proposed methodology to get an LTI model relating the longitudinal velocity with the PWM input to the brushless motors. It also explains the battery effects on the system dynamics.

3.1 Preprocess the Data

As discussed in Sect. 2, the car motion and position can be captured by the Qualisys motion system. The longitudinal velocity is given in global coordinates and transformed to body frame using the following relation:

$$v_B = R \cdot v_I, \tag{1}$$

where v_B is the linear velocity in body frame, R is the rotation matrix, and v_I is the linear velocity in inertial frame.

The *PWM* values are extracted on a SD card with a varying sampling time that varies depending on the computation time at each instant. The information from the camera and SD are captured at different frequencies, so to overcome this problem both signals are first modified as follows:

$$x_{scaled} = \frac{x_i - \mu(x_i)}{\sigma(x_i)}, \tag{2}$$

where x_i is the signal, μ is the signal's mean value, and σ its standard deviation.

Then, by tuning the index of the longitudinal velocity V_x extracted from the external camera and by interpolating it with the PWM signal using the variable time step from the microcontroller, the two signals are now synchronized.

For the upcoming sections, all the models are identified with a PWM ranging between [−500] and [500] corresponding to minimum and maximum value of [976] and [1976] on the real system, these intervals are the ones defining the operating conditions.

3.2 Longitudinal Dynamic Model

This subsection focuses on the modeling part of the longitudinal vehicle dynamics. Vehicle modeling is difficult due to its nonlinearities and uncertainties, as well as the fact that it must take into account several dynamic components that interact with one another (motors, tire model, frictions...). In [8], the authors have proposed a second-order transfer function to identify the longitudinal dynamics. In this section, a first-order system model is selected. Under the assumption of no slipping and negligible friction, the following simple linear regression allows to map the PWM to the wheel/vehicle speed:

$$V_x = p_1.PWM + p_2, \tag{3}$$

where p_1 is the slope of the estimated linear equation seen in Fig. 2 and p_2 the V_x-intercept, represents the intersection between the linear equation and the y-axis, while ensuring that $p_2 = 0$.

Now, in order to model the system dynamics, the following first-order system is proposed:

$$G(s) = \frac{K}{Ts + 1}, \tag{4}$$

where T represents the time constant of the system and K the gain represents the slope of the estimated linear regression. The time constant is estimated as $T = 0.4s$ from the main dynamics using a step response of the car.

The battery voltage and state of charge have a large effect on the vehicle longitudinal dynamics. We then carried out several experimental tests with different batteries, at different voltages. We here consider a 80% battery state of charge to get the data for system identification. After synchronizing the two data (the velocity captured

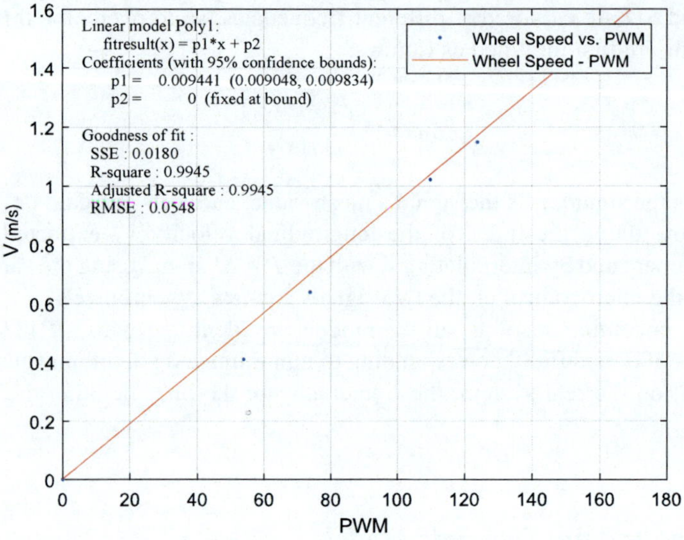

Fig. 2 Mapping wheel speed to PWM

by the cameras and the PWM logged in the SD) and extracting the linear velocity corresponding to their PWM values and referring to Figs. 2 and 3, the estimated linear model is validated, as the two slopes and linear estimations are very similar.

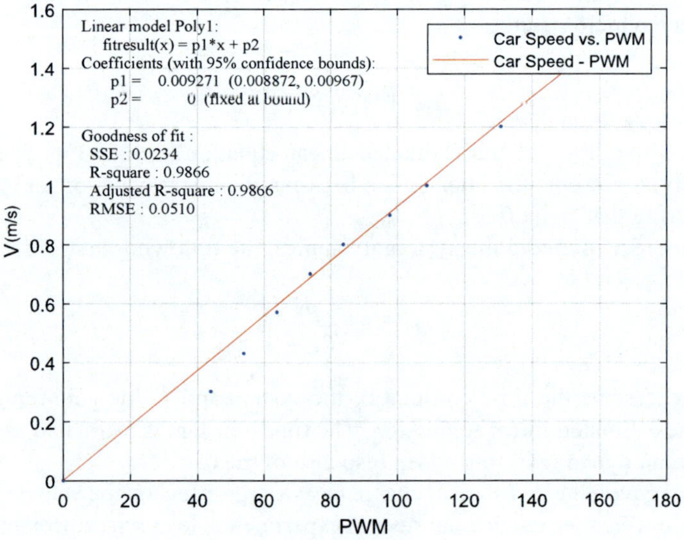

Fig. 3 Mapping car speed to PWM

4 Adaptive Cruise Control in H_∞ Control Frame

Adaptive Cruise Control system is an extension of the Cruise Control. Its goal is to control the car's longitudinal motion, such as its longitudinal velocity, acceleration, or longitudinal distance from another preceding vehicle in the same lane. The considered design method here is the H_∞ control approach for which technical details can be found in [11].

4.1 Spacing Policy

The Adaptive Cruise Control system is based on the spacing policy. It is defined as the optimal spacing that a vehicle attempts to maintain with respect to its preceding. As shown in Fig. 4, the relative inter-vehicle distance value is d_i, this value is based on the look ahead relative distance between both vehicles, it is also called velocity-dependent spacing policy.

In this paper, the constant time-headway policy is considered, to follow the leading vehicle (i) with a desired relative distance defined as:

$$d_{r,i} = d_{0,i} + hv_i, \qquad 1 \leq i \leq n \tag{5}$$

where $d_{r,i}$ is the desired safety distance between the two vehicles, represents the desired relative distance between the rear bumper of the vehicle $x_{r,i-1}$ and the front bumper of the following vehicle $x_{f,i}$, $d_{0,i}$ the standstill distance equal to $0.1m$, h is the desired time headway equal to $0.2s$, represents the time that the following vehicle needs to reach the same position as its preceding, and v_i the longitudinal velocity of the (ith) vehicle.

The relative distance is defined as follows:

$$d_i = x_{r,i-1} - x_{f,i}. \tag{6}$$

Finally, the spacing error represents the relative difference between the relative distance and the desired relative distance between vehicles, and it is defined as follows:

$$e_i = d_{r,i} - d_i = x_{r,i-1} - x_{f,i} - d_{0,i} - hv_i. \tag{7}$$

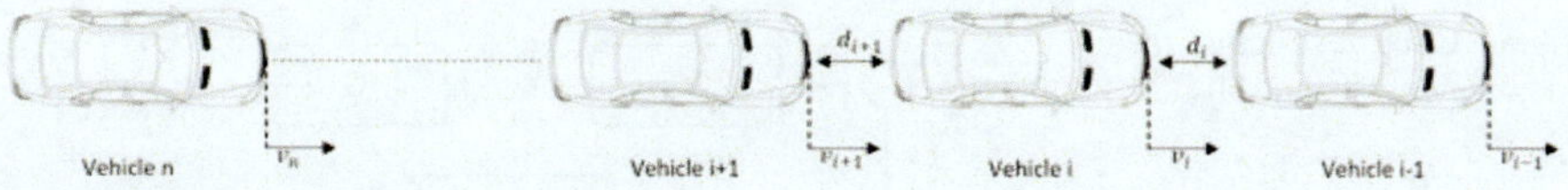

Fig. 4 Scheme of adaptive cruise control

4.2 CC and ACC Problems Definition

The Cruise Control and Adaptive Cruise Control problems are represented as in Figs. 5 and 6 below.

The lower level controller K is responsible to regulate the throttle-accelerator to correctly maintain a desired speed chosen by the driver without any outside interference in the CC framework. However, in the ACC, its objective is to follow the preceding vehicle at a desired relative distance while ensuring the inter-vehicle safety distance. In a real car, this distance is calculated using a look forward sensors, i.e., radar, LiDAR.

In the CC/ACC framework, the aim is to ensure a good tracking with zero steady-state error, fast rise time, small overshoot, and above all, the comfort and maintaining the stability of the system. The aim of this controller is to ensure a convergence towards zero of the spacing error (7), which expresses the difference between the relative distance and the desired relative distance. To meet that requirement, an integrator is added at the output of the longitudinal model (as the longitudinal dynamical model of the vehicle (4) relates the longitudinal velocity to the PWM input) in order to get the instant position of the vehicle, as seen in Fig. 6.

Finally, the transfer function of the velocity-dependent spacing policy is written as

$$H_i(s) = hs + 1 \qquad 1 \le i \le n. \tag{8}$$

4.3 CC and ACC H_∞ Control Problems

To reach the objectives an H_∞ approach is used to control the longitudinal motion of the vehicle. The aim of this approach is to synthesize a controller K that minimizes the H_∞ norm of the closed-loop system (between the external inputs w to the controlled outputs z), as

$$||T_{wz}(s)||_\infty \le \gamma, \tag{9}$$

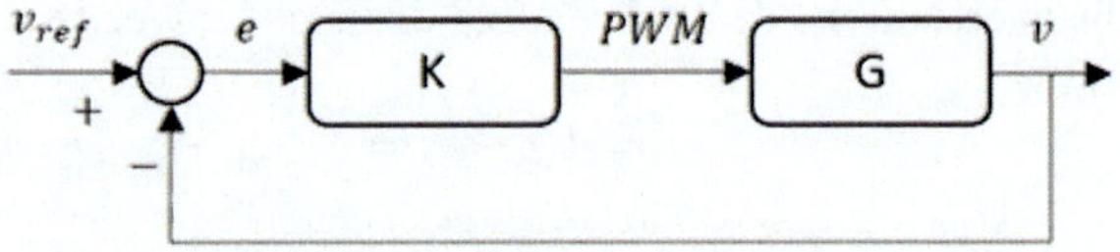

Fig. 5 Control structure for cruise control

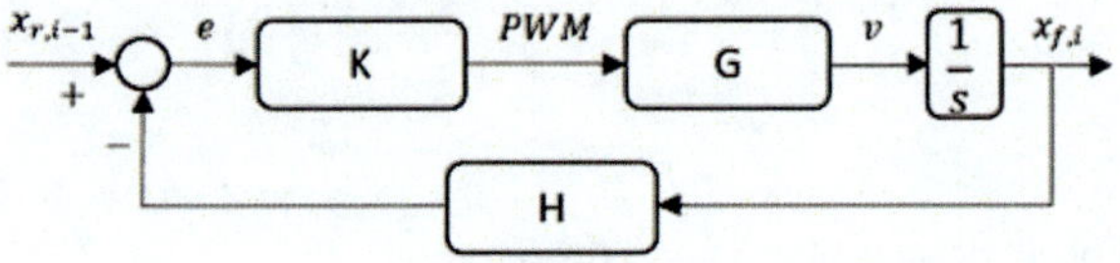

Fig. 6 Control structure for adaptive cruise control

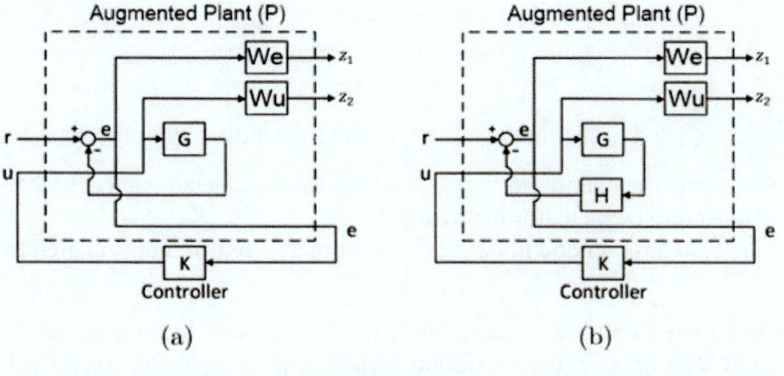

Fig. 7 Closed-loop system for CC (**a**) and ACC (**b**)

where γ is the best-achieved H_∞ norm of the closed-loop system $T_{wz}(s)$.

The performance specifications are formulated through weighting functions on the closed-loop system. We will follow here a method similar to [12] for the ACC case, with slight modifications on the weighting functions. Therefore, we have selected W_e for the tracking error and W_u for the actuator limitations, as given in the new control scheme as shown in Fig. 7a and b for Cruise Control and Adaptive Cruise Control, respectively.

The augmented plant P encompasses the model of the system to be controlled plus the performance weights, which has two inputs and three outputs as shown in Fig. 7a and b: r the exogenous input, u the control input of the plant, z_1 and z_2 the two exogenous outputs, and e the controller input.

The first weighting function, W_e, is chosen to ensure a good robustness margin, small steady-state error, and fast tracking of the reference with respects to actuators time response, which is represented by

$$W_e = \frac{\dfrac{s}{M_s} + \omega_b}{s + \omega_b \varepsilon}, \tag{10}$$

where M_s, ω_b, and ε parameters to be tuned to meet the requirements as follows:

- $M_s = 2$, to ensure sufficient module margin,
- $\omega_b = 8$ rads, to ensure a fast tracking of the reference with respect to the actuator bandwidth,
- $\varepsilon \leq 1e^{-4}$, to have a small steady-state error.

On the other hand, the second weighting functions, W_u, are chosen in a way to minimize the actuator control effort, while ensuring a good noise rejections from the inputs and fixing the maximum allowed variations control efforts of the actuators, they are represented as follows:

$$W_u = \frac{s + \dfrac{\omega_{bc}}{M_u}}{\varepsilon_1 s + \omega_{bc}}, \tag{11}$$

where M_u, ω_{bc}, and ε_1 parameters to be tuned to meet the requirements as follows:

- M_u represents the maximum variation of the reference with respect to the input,
- ω_{bc} is related to the actuator bandwidth,
- $\varepsilon_1 \leq 1e^{-2}$, to have a good noise rejection from the control inputs at high frequencies.

For the Cruise Control case, the parameter M_u represents the maximum variations of the input (PWM) over the maximum variations of the reference (v_{ref}), it is equal to 250. In order to meet the new requirements, as in the ACC the reference is the leading vehicle position, the parameter M_u is modified and it is equal to 625. The parameter ε in the error weighting function is tuned as well, to have a very low steady-state error at low frequency, it is less or equal to $1e^{-5}$.

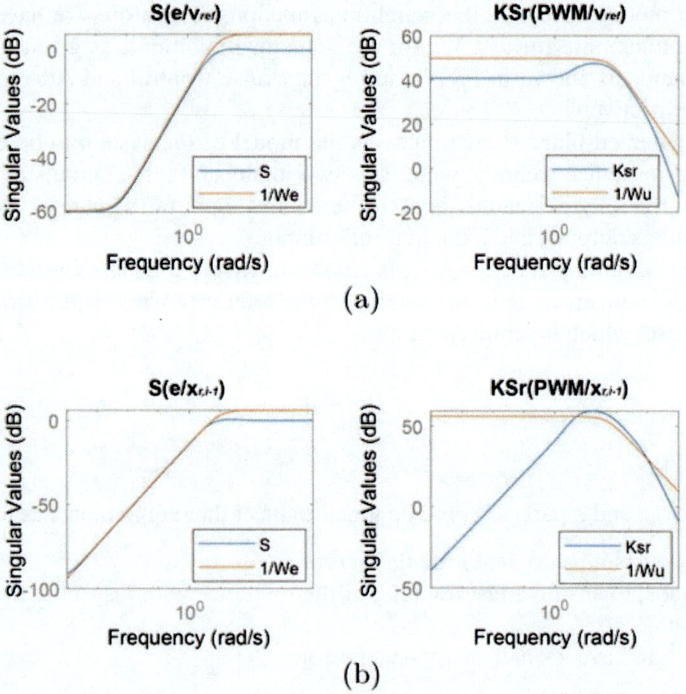

Fig. 8 Frequency domain response of the H_∞ control for the CC (**a**) and ACC (**b**)

4.4 Frequency Domain Analysis for CC and ACC

The controller performances are now analyzed in frequency domain in order to determine whether the controllers satisfy the requirements or not. The input/output performances are defined by different sensitivity functions as shown in Fig. 8a and b. It can be seen that the sensitivity functions $S = \frac{e}{v_{ref}}$ and $KSr = \frac{PWM}{v_{ref}}$ globally meet the requirements since they are below the templates with $\gamma = 0.78$ for Cruise Control and, with a very small overpass of the template for KSr in Adaptive Cruise Control (where $\gamma = 1.48$). The control input analysis is carried out to study the behavior of the controllers at low and high frequencies. Referring to KSr shown in Fig. 8a and b, there is a low risk of actuators saturation and good noise attenuation at the control input. The different sensitivity functions meet the requirements with a small steady state error, good noise, and disturbance rejections while ensuring a good module margin and fast time response of the system, despite the small overpass shown in the controller sensitivity KS in ACC case.

5 Simulation Results

The Matlab simulations are carried out using the LTI first-order longitudinal dynamical model (4) in discrete-time domain with a sampling time $T_s = 0.03s$.

Remark 1 The sampling time is chosen based on the mean value of the variable time step exported directly from the microcontroller.

To test the longitudinal control performance for the Stop & Go mode which includes the Cruise Control (CC) and the Adaptive Cruise Control (ACC) systems, using the LTI controllers calculated in Sect. 4.3, two scenarios are considered as follows:

Scenario 1: Test the Cruise Control system
Referring to Fig. 9, as emphasized before in the frequency-domain analysis the LTI controller satisfies the requirements with a small steady-state error and fast time response without any overshoot. Taking into consideration the same generated reference velocity, the adaptive cruise control system is tested in the second scenario.

Scenario 2: Follow the preceding vehicle starting from stationary state and with relative distance equal to 0.1 m (ACC).
The leading vehicle is simulated using the CC system, and the following vehicle using the ACC system in order to follow its preceding at an inter-vehicle safety distance. Referring to Fig. 10a, the proposed controller showed satisfactory results since there is a fast tracking to the generated reference, with a very small steady-state error and no overshoot. As discussed in the Sect. 4.1, the spacing policy is called velocity-dependent spacing policy as well, since its transfer function

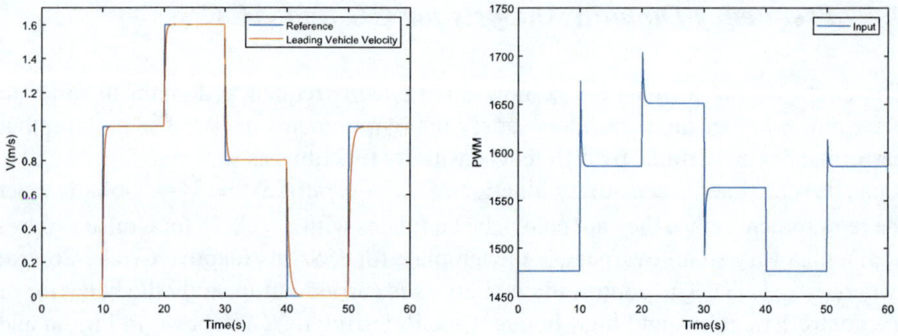

(a) Time response of speed for CC. (b) The corresponding system input.

Fig. 9 Simulation of the cruise control system

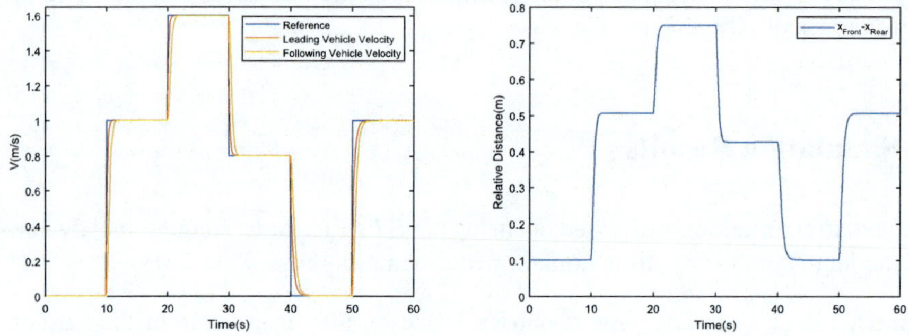

(a) Time response of speed for ACC. (b) The relative distance between vehicles.

Fig. 10 Simulation results for ACC with square leading vehicle velocity profile

(8) depends on the following vehicle velocity. The relative safety distance is represented in Fig. 10b, it is clear that for each vehicle speed there is a different inter-vehicle distance.

The next section will address the implementation of the developed controllers on a real scaled vehicle.

6 Experimental Results

To assess the performances of the proposed longitudinal control methods, some experimental tests have been carried out on the real scaled car.

After implementing the discrete identified model with its relevant H_∞ controller for the Stop&Go scenario on the pixhawk microcontroller board, the cruise control system with the signs detection is tested in a zero-loop test-track. This test-track is

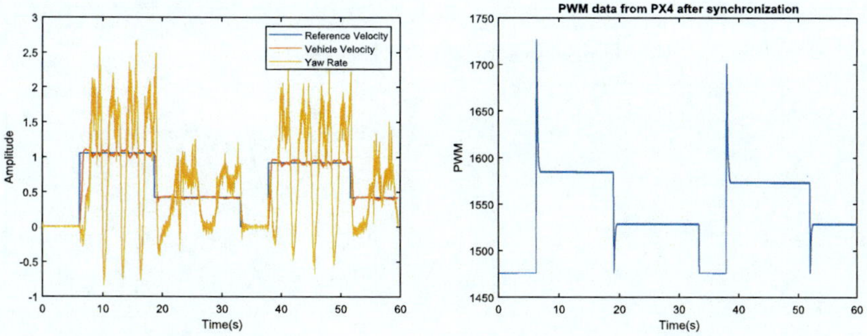

(a) Time domain response of CC. (*Yaw Rate (rad/s), Velocity(m/s)*)

(b) The corresponding System input after synchronization.

Fig. 11 CC System's experimental results for different velocities generated by the "road signs" detection

challenging due to the different existing scenarios: slowing down, speeding up, staying between the center lanes using the lateral control, which the car came equipped with.

As a result, Fig. 11a and b corresponds to the first scenario (speed tracking). The proposed H_∞ controller shows satisfactory results with fast tracking performance, without any overshoot and zero steady-state error. These shown oscillations are due to the two sharp turns in the track (which induces a coupling between longitudinal and lateral dynamics), and to the conversion from global frame to local frame in addition to the external camera data noises.

Now, for the second scenario, corresponding to the Stop&Go mode, the ACC system is tested on a straight lane (to avoid any oscillation due to the lateral motion and for the sake of clarity and accuracy). Referring to Fig. 12a–d, for the reference velocity $v_{ref} = 0.55\,\mathrm{m/s}$ and $v_{ref} = 0.8\,\mathrm{m/s}$, respectively, there is a fast tracking of the preceding vehicle with a coherent relative inter-vehicle distance, despite the small overshoot in the first case. The proposed H_∞ controllers show very promising results while satisfying the requirements cited before.

7 Conclusions

In this paper, we have proposed and experimentally validated the longitudinal control of autonomous vehicles in CC and ACC using an H_∞ approach. The proposed controllers meet the physical specifications and limitations, thanks to the tuning parameters set in the frequency domain.

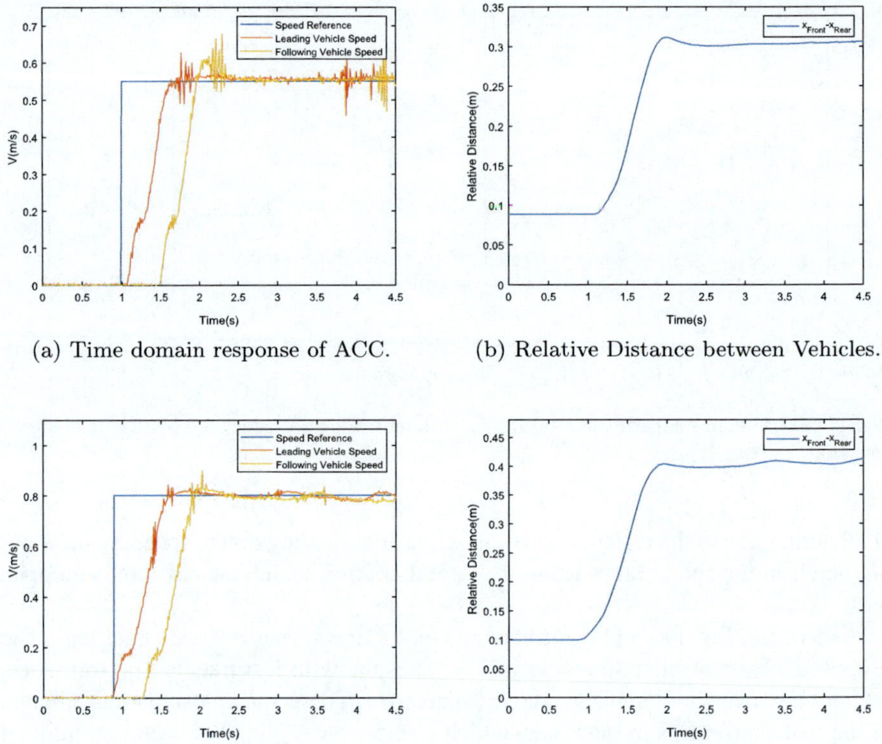

(a) Time domain response of ACC. (b) Relative Distance between Vehicles.

(c) Time domain response of ACC. (d) Relative Distance between Vehicles.

Fig. 12 Experimental validation of the ACC by simulating a virtual leading car using the CC

The comparison revealed that the experimental controller's behavior for the ACC problem is consistent with theory. Both controllers presented were able to handle multiple objectives starting safety, speed regulation, and fast time response.

Future works may concern the use and validation of an LPV control approach to handle the variations of the time-headway online set by the driver, in particular in the Cooperative Adaptive Cruise Control $CACC$ case, which takes into consideration the inter-communication delay between vehicles [9].

References

1. Ploeg J, Shukla DP, van de Wouw N, Nijmeijer H (2014) H_∞ Controller synthesis for string stability of vehicle platoons. IEEE Trans Intell Transp Syst 15(2):854–865
2. Wang C, Nijmeijer H (2015) H_∞ String stable heterogeneous vehicle platoon using cooperative adaptive cruise control. In: 2015 IEEE 18th international conference on intelligent transportation systems (ITSC). IEEE, pp 1977-1982

3. Németh B, Gáspár P (2013) Design of vehicle cruise control using road inclinations. Int J Veh Auton Syst 11(4):313–333
4. Németh B, Bede Z, Gáspár P (2017) Control design of traffic flow using look-ahead vehicles to increase energy efficiency. In: American control conference (ACC), Seattle, USA, pp 3530–3535
5. Mihály A, Németh B, Gáspár P (2014) Look-ahead control of road vehicles for safety and economy purposes. In: European control conference (ECC). Strasbourg, France, pp 714–719(2014)
6. Xiao L, Gao F (2010) A comprehensive review of the development of adaptive cruise control systems. Veh Syst Dyn 48:1167–1192
7. Li SE, Li K, Rajamani R, Wang J (2011) Model predictive multi-objective vehicular adaptive cruise control. IEEE Trans Control Syst Technol 19(3):556–566
8. Flores C, Milanés V (2018) Fractional-order-based ACC/CACC algorithm for improving string stability. Transp Res Part C: Emer Technol 95:381–393
9. Laib K, Sename O, Dugard L (2020) String stable H_∞ LPV cooperative adaptive cruise control with a variable time headway. IFAC-PapersOnLine 53(2):15140–15145
10. Tsai C-C, Hsieh S-M, Chen C-T (2010) Fuzzy longitudinal controller design and experimentation for adaptive cruise control and Stop & Go. J Intell Robot Syst 59:167–189
11. Skogestad S, Postlethwaite I (2005) Multivariable Feedback Control. Analysis and Design. Wiley, Chichester
12. Kayacan E (2017) Multiobjective H_∞ control for string stability of cooperative adaptive cruise control systems. IEEE Trans Intell Veh 2(1):52–61

Decoupled Control of a Twin Hull-Based Unmanned Surface Vehicle Using a Linear Parameter Varying Approach

Echrak Chnib, Olivier Sename, Francesco Ferrante, Eduardo S. Rodriguez Canales, and Juan C. Cutipa Luque

Abstract This paper presents a decoupled control of a twin hull-based Unmanned Surface Vehicle EDSON-J, using the H_∞ approach for Linear Parameter Varying (LPV) polytopic systems. This method was adapted in order to guarantee robust stability and performance regarding the important nonlinearities and the uncertainties on the hydrodynamics parameters. After the presentation of the nonlinear model and the decoupled model of the USV considered for the study, an LPV model was built, regarding the mass of the vehicle as unique varying parameter. Then the methodology of the control law applied is exposed and simulation results are presented. A comparison with the LTI/H_∞ approach will show the interest of the method in terms of performance.

Keywords Linear parameter varying (LPV) · H_∞ robust control · Unmanned surface vehicle (USV) · Decoupled control

E. Chnib (✉) · O. Sename · F. Ferrante
University Grenoble Alpes, CNRS, Grenoble INP, GIPSA-Lab 38000 Grenoble, France
e-mail: echrak.chnib@grenoble-inp.fr

O. Sename
e-mail: olivier.sename@grenoble-inp.fr

F. Ferrante
e-mail: francesco.ferrante@gipsa-lab.fr

E. S. Rodriguez Canales · J. C. Cutipa Luque
Electronic Engineering Department, Universidad Nacional de San Agustín de Arequipa, Arequipa, Peru
e-mail: erodriguezca@unsa.edu.pe

J. C. Cutipa Luque
e-mail: jcutipalu@unsa.edu.pe

G. Conte and O. Sename (eds.), *Proceedings of 10th International Conference on Mechatronics and Control Engineering*, Lecture Notes in Mechanical Engineering,
https://doi.org/10.1007/978-981-19-1540-6_10

1 Introduction

Since two thirds of the Earth's surface is covered by the oceans, the development of water vehicles has been seeing an increasing interest [3]. Over the past decades, several works on Unmanned Maritime Vehicles (UMVs) have been developed comprising unmanned surface vehicles (USVs) and unmanned undersea vehicles (UUVs). As global positioning systems have become more compact, the use of unmanned surface vehicles (USVs), also known as autonomous surface crafts (ASCs), in performing complex tasks has become more and more popular [9]. Applications in which USVs are used include mine counter measures (MCM), inspections for inshore environmental monitoring [10]. An overview on the research and applications of USVs is presented in [7].

Due to cross-coupled dynamics and the presence of hydro-dynamic forces, the design of efficient control systems for UMVs is generally challenging. In [13], a nonlinear gain-scheduling control was proposed to control the INFANTE AUV. A diving-control design, based on Lyapunov theory and backstepping techniques, is proposed to control an AUV in [5]. In [2], a robust H_∞ control approach, based on a two-degrees-of-freedom controller, was developed for the control of AUV. Recently, a nonlinear control scheme based on the energy-shaping (ES) principle and state error port-controlled Hamiltonian (PCH) systems has been proposed in [8]. The design of robust adaptive steering controllers was proposed in [15] to deal with the control of USVs in the presence of uncertainties, unknown control direction, and input saturation.

For systems with predominant nonlinear behaviors and large parameter uncertainties, the use of robust control techniques, such as H_∞ control, may not be sufficient to achieve good performance over the full range of parameter variation. In this setting, a possible approach to guarantee robustness and performance consists of relying on Linear Parameter Varying (LPV) polytopic models and H_∞ control; see, e.g., [11]. In the last years, the use of LPV controllers has shown to be effective in the control of autonomous surface craft. In [6], the authors presented an LPV controller with adaptive parameter estimation and adaptation capabilities. More recently, in [12] an LPV robust control system for a twin hull-based unmanned surface vehicle, named EDSON-J, was developed to account for mass variation considering a set of controllers generated by using the so-called grid-based approach.

In this paper, we aim to extend the design of LPV controllers for EDSON-J model by considering a decoupled control approach. In particular, due to the inherent complexity in embedding the LPV controller proposed [12] on a low-cost platform, the major contribution of this paper consists of designing two independent controllers for both lateral motion and longitudinal motion guaranteeing performance and robustness. This leads to controllers that can be easily implemented in low-cost embedded systems, thereby making our results more appealing in practice.

The paper is organized as follows. Section 2 presents the EDSON-J architecture. The decoupled model into surge–yaw of the USV is considered in Sect. 3.

Section 4 presents the proposed H_∞/LTI surge–yaw decoupled control. The LPV surge–yaw decoupled control is developed in Sect. 5. Section 6 discusses the obtained simulation results followed by conclusions in Sect. 7.

2 EDSON-J Architecture

The EDSON-J is a twin hull-based Unmanned Surface Vehicle that has been developed at the University of San Agustin de Arequipa, Peru, in 2019. This autonomous vessel is designed to perform inspection and monitoring tasks in the Pacific sea coastal and lagoons of Arequipa. Table 1 summarizes the main mechanical specifications of this USV. The vehicle design consists of two slender bodies propelled by two electric motors in differential mode n_d and common mode n_c. The EDSON-J's operating speed is 1.5 m/s with a maximum speed of 2.5 m/s. The vessel power supply is provided by two 100 Ah AGM batteries. The navigation system of the USV is based on a dual GNSS/INS system in order to ensure the position accuracy, velocity, and acceleration measures. The communication system is provided by RF transceivers, with a range of 96.56 km in the 900 MHz spectrum. The control architecture of the vehicle is consisting of an embedded computing board with an operative robotic system (ROS) running on GNU/Linux for the implementation of the control algorithms.

2.1 *Nonlinear Model of the EDSON-J*

According to the nomenclature given in [4], the mathematical model of the surface vehicle has been developed by considering the interaction between the rigid body dynamics (hull structure) and hydrodynamics forces interaction (the fluid). The Fig. 1

Table 1 EDSON-J USV dimensions

Parameter	Value
Length	3.00 m
Breadth	1.60 m
Mass	250 kg
Payload	100 kg
Hull breadth	0.30 m
Draft	0.20 m
Moment of inertia	201.1 kg/m^2
Location of mass center	0.11 m
Electric motor power (each)	600 W

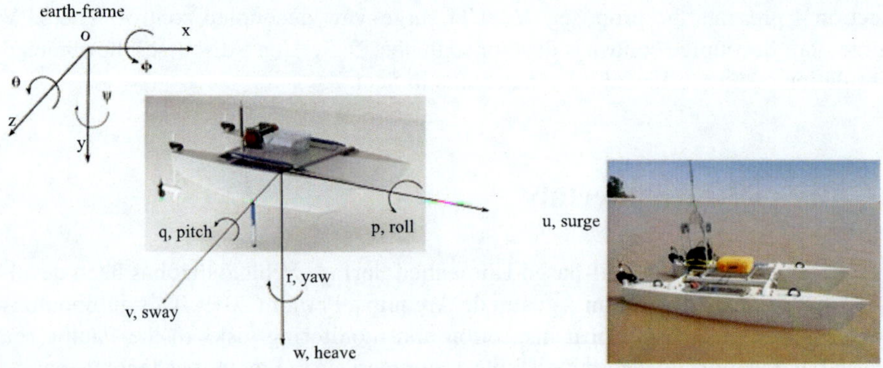

Fig. 1 Coordinate frames of the EDSON-J [12]

presents the earth-frame and the rigid body-frame fixed at the centroid of the USV. The inertial frame is represented by the position vector $\eta = \left[x\ y\ \psi\right]^T$, where x is the surge position, y is the sway position, and ψ is the yaw position. The body-frame is represented by the velocity vector $\nu = \left[u\ v\ r\right]^T$, where u is the linear velocity of the surge, v is the linear velocity of the sway, and r is the angular velocity of the yaw. Then the nonlinear model of the EDSON-J can be represented by the kinematic and dynamic equations, respectively (1) and (2):

$$\dot{\eta} = J(\eta)\nu \tag{1}$$

$$M\dot{\nu} + C(\nu)\nu + D(\nu)\nu = \tau \tag{2}$$

where $\dot{\eta}$ stands for the derivative of the position vector η and $J(\eta)\nu$ the coordinate transformation matrix from the earth-frame to the body-frame defined as

$$J(\eta) = \begin{bmatrix} \cos(\psi) & -\sin(\psi) & 0 \\ \sin(\psi) & \cos(\psi) & 0 \\ 0 & 0 & 1 \end{bmatrix} \tag{3}$$

$\tau := \left[X\ Y\ N\right]^T$ is the vector of the generalized input forces originated by the propeller actuators. X is the force in surge direction, Y is the force in sway direction, and N is the moment in yaw direction. The dynamic equation is defined by its inertia matrix M, the C matrix containing Coriolis and centrifugal forces, and the damping matrix D. These matrices are expressed as follows:

$$M = \begin{bmatrix} m - X_{\dot{u}} & 0 & 0 \\ 0 & m - Y_{\dot{v}} & mx_g - Y_{\dot{r}} \\ 0 & mx_g - N_{\dot{v}} & I_z - N_{\dot{r}} \end{bmatrix} \tag{4}$$

Table 2 EDSON-J USV parameters

Parameter	Value	Unit	Parameter	Value	Unit
$X_{\dot{u}}$	–2.471	Kg	X_u	–0.2912	Kg/s
X_{uu}	–27,626	Kg/m	X_{un_c}	–3.682	Kg
X_{rn_d}	2.762	Kg	$X_{n_cn_c}$	1.052	Kg/m^2
$X_{n_dn_d}$	1.052	Kg/m^2	$Y_{\dot{v}}$	–247.065	Kg
$Y_{\dot{r}}$	–370.597	Kg.m/rad	Y_v	–164.71	Kg/s
Y_{vv}	–38.928	Kg/m	Y_{vn_c}	–0.359	Kg
Y_{rn_c}	–0.538	Kg.m	$N_{\dot{r}}$	–748.310	Kg.m^2/rad
$N_{\dot{v}}$	–370.597	Kg.m	N_r	–988.260	Kg.m^2/rad.s^2
N_{rr}	–262.791	Kg.m^2.rad^2	N_{un_d}	2.762	Kg.m
N_{rn_c}	–2.855	Kg.m^2	$N_{n_Cn_d}$	–1.578	Kg.m^2
N_{vn_c}	–0.538	Kg.m			

$$C = \begin{bmatrix} 0 & -mr & -mx_g r + Y_{\dot{v}} v + Y_{\dot{r}} r \\ mr & 0 & -X_{\dot{u}} u \\ mx_g r - Y_{\dot{v}} v - Y_{\dot{r}} r & X_{\dot{u}} u & 0 \end{bmatrix} \tag{5}$$

$$D = \begin{bmatrix} X_u + X_{|u|u} |u| & 0 & 0 \\ 0 & Y_v + Y_{|v|v} |v| & 0 \\ 0 & 0 & N_r + N_{|r|r} |r| \end{bmatrix} \tag{6}$$

where m is the USV mass, x_g the position of center of mass and I_z the moment of inertia around the z axis, $X_{\dot{u}}$, $Y_{\dot{v}}$, $Y_{\dot{r}}$, $N_{\dot{v}}$, and $N_{\dot{r}}$ are the hydrodynamic terms related to added masses. The terms X_u, Y, and N_r represent the linear drag coefficients, while $X_{|u|u}$, $Y_{|v|v}$, and $N_{|r|r}$ are the quadratic drag coefficients [12].

2.2 *Hydrodynamic Coefficients*

The vehicle parameters can be obtained through either empirical relations or numerical/computational methods using the slender-body theory and the rigid body dynamics. The parameters of the USV nonlinear model are reported in Table 2. More details about the hydrodynamic coefficients can be found in [12].

3 Decoupled Models of the EDSON-J

This section presents the decoupled model of the EDSON-J into two subsystems: the surge and the sway/yaw. From the nonlinear model of the EDSON-J expressed by (1) and (2), we derive a linearized model of the following form:

$$G : \begin{cases} \dot{x} = Ax + B\chi \\ y = Cx + D\chi \end{cases} \tag{7}$$

$x = \begin{bmatrix} u \; v \; r \end{bmatrix}^T$ is the state vector, $\chi = \begin{bmatrix} n_c \; n_d \end{bmatrix}^T$ is the control input vector, and $y = \begin{bmatrix} u \; r \end{bmatrix}^T$ is the output vector. Matrices A, B, C, and D are obtained by Jacobian linearization around the cruise speed $u_0 = 2$ m/s and a rotational speed $n_0 = 14$ rps.

3.1 Surge and Sway/Yaw LTI Models

Surge LTI Model The surge system corresponds to the dynamics of the state u in (7), with control input n_c and measured output $y_s = u$. A state space representation of this system is given next, with $x_s = u$:

$$S : \begin{cases} \dot{x}_s = A_s x_s + B_s n_c \\ y_s = C_s x_s + D_s n_c \end{cases} \tag{8}$$

$$A_s = \frac{-X_u - X_{un_c} n_0 - 2X_{uu} u_0}{X_{\dot{u}} - m}, \quad B_s = \frac{-2X_{n_c n_c} n_0 - X_{un_c} u_0}{X_{\dot{u}} - m} \tag{9}$$

$$C_s = 1, \quad D_s = 0 \tag{10}$$

Sway/Yaw LTI Model The sway/Yaw system corresponds to the dynamics of the state vector $x_{sy} = \begin{bmatrix} v \; r \end{bmatrix}^T$ in (7), the control input n_d and the measured output vector $y_{sy} = r$. It can be represented by the following state space equations:

$$SY : \begin{cases} \dot{x}_{sy} = A_{sy} x_{sy} + B_{sy} n_d \\ y_{sy} = C_{sy} x_{sy} + D_{sy} n_c \end{cases} \tag{11}$$

3.2 Surge and Sway/Yaw LPV Models

The above models are LTI models if the mass m is known and fixed. As opposed, when the mass takes value in $[m_{min}, m_{max}]$, those models can be represented as LPV systems; see, e.g., [12], wherein a similar approach is pursued. **Surge LPV Model** As seen in (9), the matrices A_s and B_s depend on m in a nonlinear fashion. To obtain a control-oriented LPV model, we can set

$$\rho_s = \frac{1}{X_{\dot{u}} - m} \tag{12}$$

which enable to obtain the following LPV model for the surge dynamics:

$$S(\rho_s): \begin{cases} \dot{x}_s = A_s(\rho_s)x_s + B_s(\rho_s)n_c \\ y_s = C_s x_s + D_s n_c \end{cases} \tag{13}$$

note that A_s and B_s are affine in ρ_s. Such a model can be easily written in a polytopic form, thereby allowing to simplify the design of a controller as detailed next.

Sway/Yaw LPV Model The matrices A_{sy} and B_{sy} also depend on m in a nonlinear fashion. However, since the mass appears both in the numerator and denominator of the entries of the matrices A_{sy} and B_{sy}, to get an LPV representation, one needs to introduce two parameters. In particular, we define

$$\rho_{sy}^1 := \frac{1}{(I_z Y_{\dot{v}} + m^2 x_g^2 - N_{\dot{r}} Y_{\dot{v}} - m I_z + m N_{\dot{r}} + Y_{\dot{r}}^2 - 2m x_g Y_{\dot{r}})} \tag{14}$$

$$\rho_{sy}^2 := \frac{m}{(I_z Y_{\dot{v}} + m^2 x_g^2 - N_{\dot{r}} Y_{\dot{v}} - m I_z + m N_{\dot{r}} + Y_{\dot{r}}^2 - 2m x_g Y_{\dot{r}})} \tag{15}$$

Following this approach, the Sway/Yaw LPV model writes:

$$SY(\rho_{sy}): \begin{cases} \dot{x}_{sy} = A_{sy}(\rho_{sy})x_{sy} + B_{sy}(\rho_{sy})n_d \\ y_{sy} = C_{sy} x_{sy} + D_{sy} n_d \end{cases} \tag{16}$$

where A_{sy} and B_{sy} are affine in $\rho_{sy} \in \mathbb{R}^2$.

4 H_∞/LTI Surge–Yaw Decoupled Control

The H_∞ control problem consists in finding a control that aims at minimizing the H_∞ norm of the closed-loop system [14]. Defining w the exogenous input vector and e the controlled output vector, the H_∞ suboptimal control problem is, given γ a predefined attenuation level, to design a controller that internally stabilizes the closed-loop system and ensures:

$$\|T_{ew}\|_\infty \leq \gamma \tag{17}$$

where T_{ew} is the closed-loop transfer matrix from w to e. The minimal value γ_{opt} is then approached by a bisection algorithm. The considered H_∞ control scheme for the two problems is chosen as in Fig. 2.

Such an H_∞ control problem can be efficiently solved by relying either on Riccati equations or linear matrix inequalities. This leads to the design of the control law u_K that solves the suboptimal problem (17) (or optimal when γ is minimized); (see [14]).

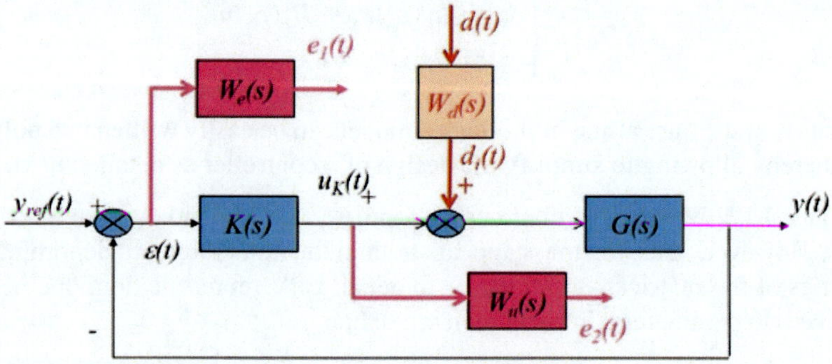

Fig. 2 Structure chosen for the Surge/Yaw controller design

4.1 Surge–Yaw Decoupled H_∞ Control

In this section, we present the structure and weighting functions selected for each decoupled subsystem.

Performance Specifications for the Surge–Yaw The weights on the sensitivity function W_{e_u} in surge velocity u and W_{e_r} yaw rate r are expressed by

$$\frac{1}{W_{e_u}} = \frac{s + \omega_{bu}\epsilon_{1u}}{\frac{s}{M_{eu}} + \omega_{bu}} \quad , \quad \frac{1}{W_{e_r}} = \frac{s + \omega_{br}\epsilon_{1r}}{\frac{s}{M_{er}} + \omega_{br}} \tag{18}$$

- For the surge:
  $M_{eu} = 2$ to ensure good robustness margin, $\epsilon_{1u} = 0.001$ to ensure a tracking error less then 1%,$\omega_{bu} = 0.46$ to ensure a fast closed-loop tracking response and rejection of disturbances.
- For the yaw:
  M_{er} =2, ϵ_{1r} = 0.001 and $\omega_{br} = 0.5$.

The weights on the controller sensitivity function W_{u_u} in surge velocity u and W_{u_r} yaw rate r are expressed by

$$\frac{1}{W_{u_u}} = \frac{\epsilon_{2u} s + \omega_{bc_u}}{s + \frac{\omega_{bcu}}{M_{uu}}} \quad , \quad \frac{1}{W_{u_r}} = \frac{\epsilon_{2r} s + \omega_{bc_r}}{s + \frac{\omega_{bcr}}{M_{ur}}} \tag{19}$$

- For the surge:
  $M_{uu} = 16$ in order to account for actuator limitations, $\omega_{bcu} = 250$ and $\epsilon_{2u} = 0.001$
- For the yaw:
  $M_{ur} = 80$ $\omega_{bcr} = 250$ and $\epsilon_{2r} = 0.001$

Finally, the weighting function W_d is set to 0.2 for the surge motion, this enables to handle the coupling with the surge dynamics, while it is set to 0 for the yaw control. Indeed, the yaw dynamics are practically decoupled from the surge.

Solving the above H_∞ control problems leads to $\gamma_{min} = 0.77$ for the surge motion, and $\gamma_{min} = 0.88$ for the yaw motion. This allows to conclude that both LTI controllers satisfy the performance requirements in the nominal case. Moreover, it may be shown that, using a μ-analysis, both controllers keep stability considering some mass uncertainty, mainly due to pay load variations, i.e., when $m \in [250, 400]$. However, they do not achieve robust performance, which emphasizes the need for an LPV controller to ensure the satisfaction of the performance requirements for large mass variations, in particular for the yaw controller. This is the objective of the following section.

5 LPV Surge–Yaw Decoupled Control of the EDSON-J

In this section, we develop LPV mass-dependent controllers for the surge and yaw dynamics. This is enabled by the use of the LPV models presented in Sect. 3. The formulation of the LPV control problem follows closely the one presented for the LTI case, using the same weighting functions for performance requirements.

5.1 LPV Control Problem

The considered LPV control problem for the surge and yaw dynamics are a direct extension of the ones proposed for the LTI control problems, yet applied on the LPV models. The control scheme is as in Fig. 3.

As mentioned earlier, it is worth noting that the weighting functions are the same for the LPV control problem than for the LTI one. This allows a fair comparison between both methodologies.

As explained in Sect. 3, the LPV surge and yaw models are affine with respect to the chosen parameters vector. This means that we can apply the polytopic approach to design both LPV controllers. This is the objective of the next section.

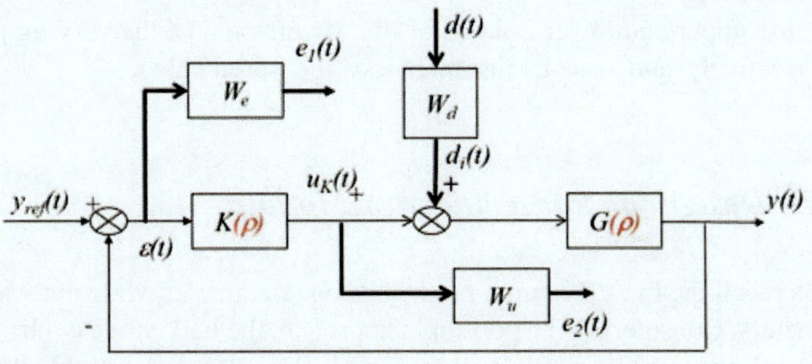

Fig. 3 LPV Surge/Yaw control design scheme

5.2 LPV Polytopic Controller

The LPV controller is designed using the polytopic approach presented in [1]. The applicability of the so-called polytopic method is restricted to LPV systems whose matrices depend in an affine fashion on the vector of parameters. More precisely,

$$A(\rho) = A_0 + \sum_{i=1}^{n_p} A_i \rho_i \tag{20}$$

In this case, the LPV system can be written as a polytopic system with 2^{n_p} vertices, where n_p is the number of the varying parameters.

In the LPV/H_∞ framework, the control synthesis problem is addressed offline by solving a set of LMIs at each vertex of the polytope using convex optimization. The parameter dependent LPV output-feedback controller is designed to guarantee the quadratic stability of the closed-loop system, together with a minimal $\mathscr{L}_2$-induced gain from the external input w to the controlled output vector e, i.e., to ensure that $||T_{ew}||_{\mathscr{L}_2-\mathscr{L}_2} < \gamma$ (where γ is to be minimized). The solution of the problem are the vertex LTI controllers, $K_i = \begin{bmatrix} A_i & B_i \\ C_i & D_i \end{bmatrix}$, where $1 \leq i \leq 2^{n_p}$. The polytopic LPV controller $K(\rho)$ is computed online as the convex combination of the vertex controllers K_i.

$$K(\rho) = \sum_{i=1}^{2^{n_p}} a_i(\rho) K_i, \text{ with } \sum_{i=1}^{2^{n_p}} a_i(\rho) = 1 \tag{21}$$

$$a_i(\rho) = \frac{\prod_{j=1}^{n_p} |\rho_j - C^c(\omega_i)_j|}{\prod_{j=1}^{n_p} (\overline{\rho_j} - \underline{\rho_j})} > 0, \quad \text{with } C^c(\omega_i)_j = \begin{cases} \overline{\rho_j} & if\ \omega_i = \underline{\rho_j} \\ \underline{\rho_j} & otherwise \end{cases} \tag{22}$$

$\overline{\rho_j}$, $\underline{\rho_j}$ are the upper and lower bounds of the jth element of the varying parameter vector, respectively, and ρ_j is its instantaneous measured value.

5.3 LPV Design for Surge and Yaw Motion

First, let us recall the the LPV surge model has one parameter while the yaw one has two parameters. Considering the polytopic approach, the LPV surge control problem includes a polytope of two vertices, while the LPV yaw control problem handles a

Table 3 H_∞ performances of the LTI and LPV closed-loop systems

	γ_{min} LTI case	γ_{min} LPV case
Surge control	0.77	0.91
Yaw control	0.88	1.22

polytope of four vertices, as can be seen later in the frequency-domain plots of the sensitivity functions.

Solving the LPV/H_∞ control problems leads to find the minimal attenuation level γ for both problems. Table 3 summarizes these results for the LPV controllers, compared with the solution of the corresponding LTI problems.

This result means that, solving the LPV yaw control problem induces some conservatism. This is due to the problem solution using a single Lyapunov function for the four vertices of the polytope. This problem can be overcome as in [12] using the gridding approach, but the latter generates a large set of controllers to be implemented, which is much more complicated to embed in a real system.

6 Simulation Results

In this section, time-domain simulation have been performed using the complete nonlinear model of the EDSON-J USV presented in [12]. Two scenario are presented to show the impact of the mass variations on the performances on both LTI and LPV control strategies.

6.1 Initial Evaluation

This first test is a basic step-response of the surge and yaw motion, here performed for the nominal ($m = 250$ kg) and worst ($m = 400$ kg) mass parameter value. As can be seen on Fig. 4, the surge motion is few affected by the difference of mass and performs well for the LTI and LPV control. However, while the yaw motion is few affected in the LPV case, it is deteriorated a lot in the LTI case for the maximal mass. This is coherent with the Robust Performance analysis carried out in Sect. 3, where we showed that the LTI yaw controller is not robust in performance.

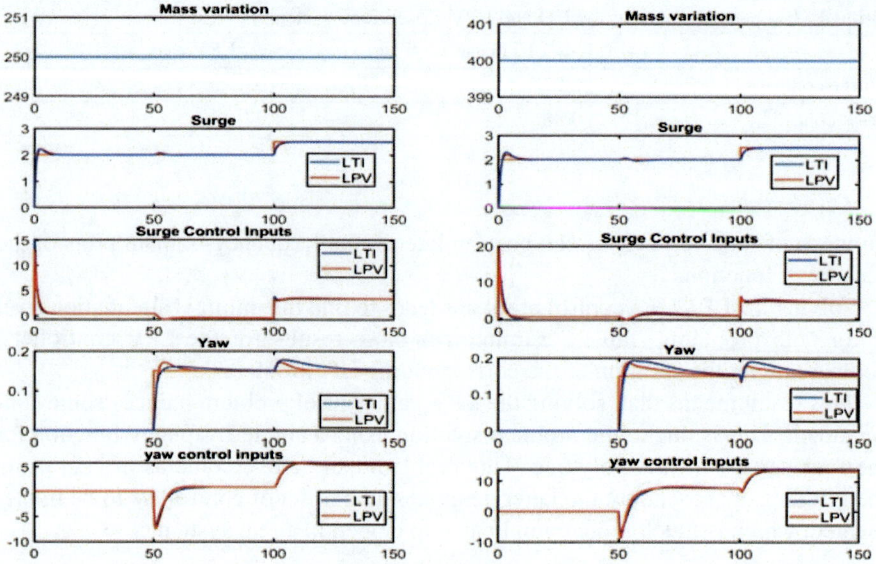

Fig. 4 Steps responses of LTI versus LPV surge/yaw controllers for the nominal (left) and maximal (right) mass value

6.2 Scenario 1

In this scenario, the surge and yaw and motion are following a series of step references. This allows to evaluate the performances of the controllers, in particular due to the coupling between longitudinal and lateral motions. As we can see on Fig. 5, the surge variable is controlled efficiently by both LTI and LPV controllers, even when the mass is varying.

However, concerning the yaw motion, it can be seen then the LTI controller is more affected by the change of mass than the LPV controller. Indeed, if we compute the root mean square error between the yaw and its reference, from $t = 150$ s (when the mass starts to change), then we get

$$RMSE(r - r_{ref})_{LPV} = 0.0130 \quad RMSE(r - r_{ref})_{LTI} = 0.0172$$

So, the LPV method improves the yaw performance by 24.4% in this scenario.

6.3 Scenario 2

In this scenario, the mass is assumed to be constant for the worst case ($m = 400$ kg), the speed motion is kept constant at $u_0 = 2$ m/s, and the yaw is following a sinusoidal

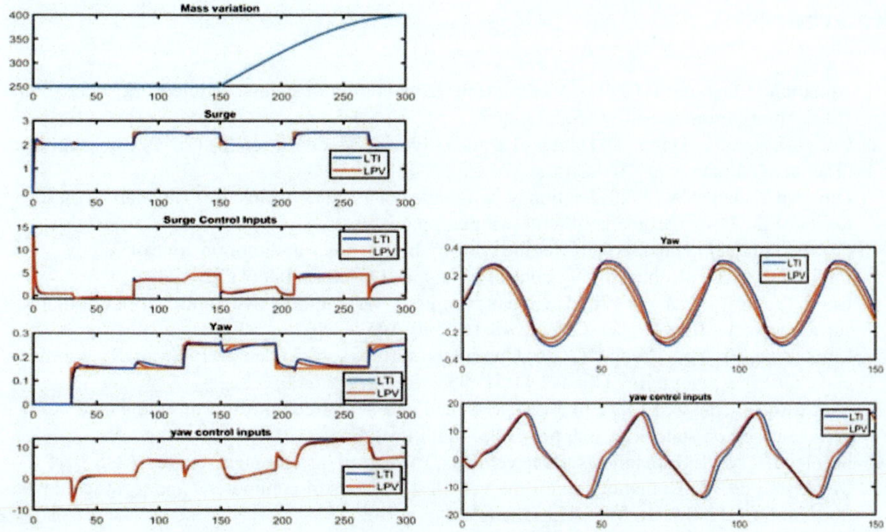

Fig. 5 LTI versus LPV decoupled control: scenario 1 (left) and scenario 2 (right)

reference between −0.25 and 0.25 rad/s. In this case, if we compute the root mean square error between the yaw and its reference then we get

$$RMSE(r - r_{ref})_{LPV} = 0.0389 \quad RMSE(r - r_{ref})_{LTI} = 0.0659$$

So, the LPV method improves the yaw performance by 40.97% in this scenario.

7 Conclusion

This paper studied the problem of robust control of the Unmanned Surface Vehicle EDSON-J. LPV mass-dependent controllers were designed in the framework of H_∞ control to fulfill performance requirements over a wide range of mass variations. The approach is deployed on a simplified model in which longitudinal and lateral dynamics are assumed to be decoupled.

Numerical simulations performed on the actual nonlinear system showed the effectiveness of the LPV control to achieve the desired performance despite the coupling between longitudinal and lateral dynamics. Comparisons with LTI controllers showed that the proposed LPV decoupled control yields better performance.

Future works may concern the design of LPV controllers to account for other parameter variations (as the hydrodynamic ones), the comparison with the global approach in [12], and the experimental validation on the real USV.

References

1. Apkarian P, Gahinet P (1995) A convex characterization of gain-scheduled $\mathscr{H}_\infty$ controllers. IEEE Trans Autom Contr 40(5):853–864
2. Cutipa-Luque JC, Donha D, Dantas JLD, de Oliveira LM, de Barros EA (2012) Robust control of an underactuated auv. IFAC Proc Vol 45(27):138–143
3. Dinstein Y, Dahl AW (2020) Section iv: unmanned maritime systems. In: Oslo manual on select topics of the law of armed conflict. Springer, pp 43–53
4. Fossen TI (2011) Handbook of marine craft hydrodynamics and motion control. Wiley
5. Lapierre L (2009) Robust diving control of an auv. Ocean Eng 36(1):92–104
6. Liu Z, Yuan C, Zhang Y (2015) Linear parameter varying adaptive control of an unmanned surface vehicle. IFAC-PapersOnLine 48(16):140–145
7. Liu Z, Zhang Y, Yu X, Yuan C (2016) Unmanned surface vehicles: an overview of developments and challenges. Annu Rev Control 41:71–93
8. Lv C, Yu H, Hua Z, Li L, Chi J (2018) Speed and heading control of an unmanned surface vehicle based on state error pch principle. Math Prob Eng
9. Manley JE (2008) Unmanned surface vehicles, 15 years of development. In OCEANS 2008:1–4
10. Manley JE (2016) Unmanned maritime vehicles, 20 years of commercial and technical evolution. In: OCEANS 2016 MTS/IEEE monterey, pp 1–6
11. Roche E, Sename O, Simon D (2009) Lpv/h_∞ control of an autonomous underwater vehicle (auv). In: Proceedings of the European control conference, pp 3160–3165
12. Rodriguez-Canales ES, Cutipa-Luque JC (2021) Lpv/h_∞ control of a twin hull-based unmanned surface vehicle. J Control, Autom Electr Syst 32(2):245–255
13. Silvestre C, Pascoal A (2004) Control of the infante auv using gain scheduled static output feedback. Control Eng Pract 12(12):1501–1509
14. Skogestad S, Postlethwaite I (2005) Multivariable feedback control. Analysis and design. Wiley, Chichester
15. Wang S, Fu M, Wang Y (2019) Robust adaptive steering control for unmanned surface vehicle with unknown control direction and input saturation. Int J Adapt Control Signal Process 33(7):1212–1224

Neural Network-Based Self-tuning Kinematic Control and Dynamic Compensation for Mobile Robots

Luis F. Recalde, Bryan S. Guevara, Danny J. Zea, and Víctor H. Andaluz

Abstract This paper proposes a neural network-based self-tuning kinematic controller and dynamic compensation for tracking trajectories, which applied, e.g., in cases where mobile robots are subject to: continuous parametric changes; different trajectories and external disturbances where online gain tuning is a desirable choice. For this kind of controller, the kinematic and dynamic model was developed considering that the mobile robot is confirmed by differential platform, and the operating point is not located at the center of wheel's axes. The artificial neural network estimates the states of the mobile robot while the gradient descent optimization algorithm adjusts the controller gains that attain the smaller position tracking error, the dynamic model is used to compensate the velocity errors in the robot. Moreover, the stability of the proposed controller is demonstrated analytically. Finally, simulation results are given considering a Turtlebot3 mobile robot, real-time experiments are implemented in the same mobile robot, where the tests are carried out to show the effectiveness of the controller in a real environment.

Keywords Neural network · Self-tuning · Optimization · Mobile robots · Auto-tuning

L. F. Recalde (✉) · B. S. Guevara · V. H. Andaluz
Universidad De Las Fuerzas Armadas ESPE, Sangolqui, Ecuador
e-mail: lfrecalde1@espe.edu.ec

B. S. Guevara
e-mail: bsguevara@espe.edu.ec

V. H. Andaluz
e-mail: vhandaluz1@espe.edu.ec

D. J. Zea
Escuela Superior Politécnica de Chimborazo ESPOCH, Riobamba, Ecuador
e-mail: danny.zea@espoch.edu.ec

G. Conte and O. Sename (eds.), *Proceedings of 10th International Conference on Mechatronics and Control Engineering*, Lecture Notes in Mechanical Engineering,
https://doi.org/10.1007/978-981-19-1540-6_11

1 Introduction

The main function of modern robotics is the development of functional robots, which are used in structured or partially structured environments, e.g. in applications like welding, cutting and exploration missions [1, 3]. Modern robotics are not only applied in the previous applications, in recent years, there has been a noticeable increase in the use of robots in industrial field, which has been one of the main precursors of more sophisticated control systems [14, 20].

The mobile robots are widely used in these applications, so these kind of systems are subject to disturbances, slippage, white noise and sensor errors, so that it is very difficult to implement high precision controllers [17]. No matter the changes in its dynamics or variations in operating conditions, the tasks must be performed with due precision, for example, in the case of load transportation, characteristics of the dynamics, center of mass and inertia are values that change when the robot is loaded. Then, to keep a good performance, the controller must be able to adapt while transporting different types of loads [17].

Some works present the design of controllers that are capable of adapting itself, e.g. [4] presents the use of a radial basis neural network (RBF-NN) for mobile robot dynamics approximation, [19] shows the use of neural networks that are based on multilayer feed forward neural networks with back-propagation learning or more efficient variations of this algorithm, researches [10, 12] present systems with a mix of neural networks and fuzzy control in which the training and rules of behavior are based on the desired state.

In this context, this paper proposed a neural network-based self-tuning kinematic controller and dynamic compensation to solve the trajectory tracking problem for a mobile robot with continuous parametric changes and different desired trajectories. The artificial neural networks are used to develop the Self-Tuning Kinematic Control, through gradient descent and back-propagation algorithm, finally the dynamic compensation, which guarantees compliance of the velocities in the real mobile robot, it is used to keep the point of interest on the desired trajectory. Simulation and real-time experiments on an Turtlebot3 mobile robot are conducted, the communication was done through ROS [11] to show the effectiveness of the proposed controller.

Including the introduction and the summary, this article is organized as follows: Sect. 2 details the kinematic and dynamic model representation of the mobile robot. The controller design and the control scheme are presented in Sect. 3. The simulation and experimental results are presented in Sect. 4. Finally, the conclusions are presented in Sect. 5.

2 General System Model of Mobile Robots

In this section, the kinematic and dynamic model of the system is presented, the system is composed by a differential mobile robot, therefore a unique and general system is formed. The configuration of a mobile platform is defined by a vector $\mathbf{q}$ with n independent coordinates, called generalized coordinates, where $\mathbf{q} = \left[q_1\ q_2\ \dots\ q_n\right]^{\mathsf{T}} = \left[\mathbf{q_p}\right]^{\mathsf{T}}$ and $\mathbf{q_p}$ represents the generalized coordinates of the mobile platform, it is appreciated that $n = n_p$, where n_p are the configuration coordinates of the mobile platform, the set of these is denoted by $\mathcal{N}$. The position and orientation of the point of interest are defined by a vector $\boldsymbol{\eta}$ given by $\boldsymbol{\eta} = \left[\eta_1\ \eta_2\ \dots\ \eta_m\right]^{\mathsf{T}} = \left[\ \boldsymbol{\eta_p}\right]^{\mathsf{T}}$ in $\mathbf{R}$ of $\boldsymbol{\eta_p}$ operational coordinates of the mobile platform, in the same way the set of these is denoted by $\mathcal{M}$ [8].

2.1 Mobile Robot Kinematic Model

Figure 1 shows a differential robot, where $\boldsymbol{\eta} = \left[x\ y\ \psi\right]^{\mathsf{T}}$ is the position and orientation of the point of interest with respect to the global reference system $< \mathcal{R} >$, where a is the distance between the geometric center and the wheel axis. The kinematic model in the mobile robot provides the derivative of the point of interest, the velocity can be represented as follows,

$$\begin{bmatrix} \dot{x} \\ \dot{y} \\ \dot{\psi} \end{bmatrix} = \begin{bmatrix} \cos(\psi) & -a\sin(\psi) \\ \sin(\psi) & a\cos(\psi) \\ 0 & 1 \end{bmatrix} \begin{bmatrix} \mu \\ \dot{\psi} \end{bmatrix} \tag{1}$$

Equation (1) can be represented in a compact form (2), where $\dot{\boldsymbol{\eta}} = \left[\dot{x}\ \dot{y}\ \dot{\psi}\right]^{\mathsf{T}}$ represents the velocities of the point of interest in the coordinate system $< \mathcal{R} >$, $\mathbf{J}(\psi)$ is the transformation matrix and $\mathbf{v}(t) = \left[\mu\ \dot{\psi}\right]^{\mathsf{T}}$ is the velocity vector of linear and angular velocities of differential robot [2].

$$\dot{\boldsymbol{\eta}}(t) = \mathbf{J}(\psi(t))\mathbf{v}(t) \tag{2}$$

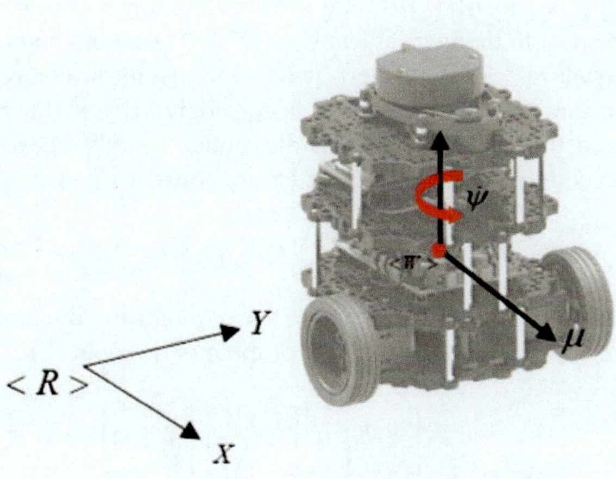

Fig. 1 Frame coordinates of differential mobile robots

2.2 Mobile Robots Dynamic Model

To obtain the dynamic model of the system, the Euler Lagrange equations are used, which propose a difference between kinetics and potential energy according to (3), and finally apply (4) [5, 15, 18].

$$\mathcal{L} = \mathcal{K} - \mathcal{P} \tag{3}$$

$$\frac{d}{dt}\left(\frac{d\mathcal{L}}{\dot{\eta}}\right) - \frac{d\mathcal{L}}{\eta} = f_i(t) \tag{4}$$

Kinetic energy is calculated from $\mathcal{K} = \frac{1}{2}\dot{\eta}^{\mathsf{T}} m\, \dot{\eta}$, and the potential energy is given by $\mathcal{P} = 0$, m represents the mass of the mobile platform, and f_i are the forces in each coordinate system, in this sense, the dynamic model shows in (5).

$$\begin{bmatrix} fx \\ fy \\ \tau\psi \end{bmatrix} = \begin{bmatrix} m & 0 & -am\sin(\psi) \\ 0 & m & am\cos(\psi) \\ -am\sin(\psi) & am\cos(\psi) & m(a^2+1) \end{bmatrix} \begin{bmatrix} \ddot{x} \\ \ddot{y} \\ \ddot{\psi} \end{bmatrix} \begin{bmatrix} 0 & 0 & -am\cos(\psi)\dot{\psi} \\ 0 & 0 & -am\sin(\psi)\dot{\psi} \\ 0 & 0 & 0 \end{bmatrix} \begin{bmatrix} \dot{x} \\ \dot{y} \\ \dot{\psi} \end{bmatrix} \tag{5}$$

The above equation can be expressed in matrix form, as

$$\mathbf{f}(t) = \mathbf{M}(\eta)\ddot{\eta}(t) + \mathbf{C}(\eta, \dot{\eta})\dot{\eta} \tag{6}$$

It is important to mention that $\mathbf{M}(\eta)$ is the inertia matrix, $\mathbf{C}(\eta, \dot{\eta})$ is the centripetal matrix and $\ddot{\eta}(t)$, $\dot{\eta}(t)$ are the acceleration and velocity vectors, respectively, with respect to the inertial frame $< \mathcal{R} >$. Turtlebot3 and others commercial robots have low level PID controllers, which does not allow controlling the motors directly using torque signals, therefore, it is considered $\mathbf{B}(\psi)$, $\mathbf{D}$ and $\mathbf{E}$ are matrices that allow the transformation of the dynamic model. Simply PD servo controllers are considered to control each motor, $\mathbf{L}$ and $\mathbf{S}$ are constants gain matrices (7).

$$\mathbf{f}(t) = \mathbf{B}(\psi)[\mathbf{D}[\mathbf{L}\mathbf{v}_{ref}(t) - \mathbf{L}\mathbf{v}(t) - \mathbf{S}\dot{\mathbf{v}}(t)] - \mathbf{E}\mathbf{v}(t)] \tag{7}$$

Finally, from Eqs. (6) and (7), we obtain the dynamic model of the mobile robot, regarding as control signals of the mobile robot (8).

$$\begin{bmatrix} \mu_{ref} \\ \psi_{ref} \end{bmatrix} = \begin{bmatrix} \chi_1 & 0 \\ 0 & \chi_2 \end{bmatrix} \begin{bmatrix} \dot{\mu} \\ \ddot{\psi} \end{bmatrix} + \begin{bmatrix} \chi_3 & \chi_4 + \chi_5\dot{\psi} \\ \chi_6\dot{\psi} & \chi_7 \end{bmatrix} \begin{bmatrix} \mu \\ \dot{\psi} \end{bmatrix} \tag{8}$$

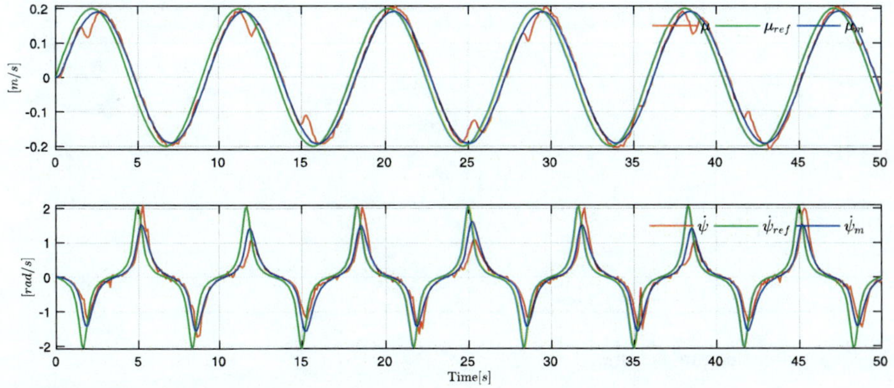

Fig. 2 Validation of the proposed dynamic model

Equation (8) can be represented in compact form (9), the dynamic model is obtained, where $\bar{\mathbf{M}}(\chi)$ represents the differential robot system's inertia, $\bar{\mathbf{C}}(\chi, \mathbf{v})$ represents the components of the centripetal forces, $\mathbf{v_{ref}}(t) = \left[\mu_{ref}\ \dot{\psi}_{ref}\right]^{\mathsf{T}}$ is a vector of reference velocities, $\mathbf{v}(t) = \left[\mu\ \dot{\psi}\right]^{\mathsf{T}}$ the real velocities of the robot, finally χ is the vector of dynamic parameters, which contain the physical parameters of differential mobile robot.

$$\mathbf{v_{ref}}(t) = \bar{\mathbf{M}}(\chi)\dot{\mathbf{v}}(t) + \bar{\mathbf{C}}(\chi, \mathbf{v}(t))\mathbf{v}(t) \tag{9}$$

The identification and validation of the proposed dynamic model are shown in Fig. 2. The identification of the mobile robot system was carried out using optimization techniques, where an objective is to minimize a cost function conformed by a vector of error between real values of the Turtlebot3 velocities and the mathematical model [16], varying the values of the vector χ.

3 Controller Design

The proposed control allows the point of interest of differential mobile robot track different trajectories. The control objective is achieved through a designed cascade control system, which comprises three stages: a kinematic controller generating the control actions for the tracking paths, the artificial neural networks are used to develop the kinematic self-tuning control, through estimation of the states of the mobile robot, the gradient descent optimization algorithm adjusts the controller gains online achieving the smallest position tracking error and a dynamic compensation that guarantees compliance with said velocities in the real robot, finally Fig. 3 presents the proposed control scheme.

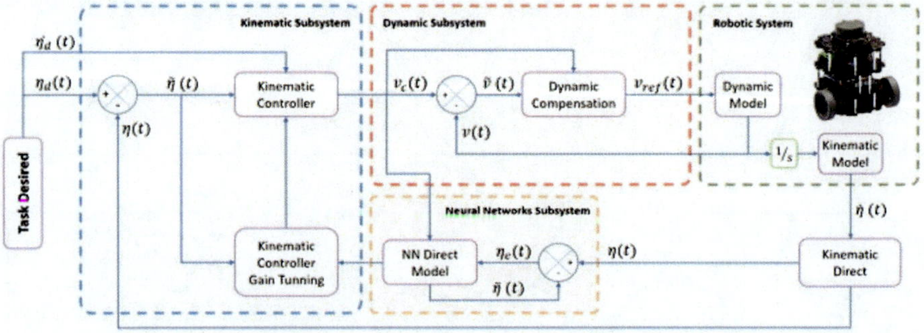

Fig. 3 Proposed control scheme

3.1 Kinematic Controller

It is proposed that inverse kinematic control (10) allows finding the velocities for the differential mobile robots that allow tracking of the desired trajectory $\boldsymbol{\eta}_\mathbf{d}(t) = \begin{bmatrix} x_d & y_d \end{bmatrix}$, where $\dot{\boldsymbol{\eta}}_\mathbf{d}(t) = \begin{bmatrix} \dot{x}_d & \dot{y}_d \end{bmatrix}$ are the velocities of the desired trajectory, $\tilde{\eta}(t) = \begin{bmatrix} \tilde{\eta}_x & \tilde{\eta}_y \end{bmatrix}$ are the control errors that are obtained from $\tilde{\eta}(t) = \boldsymbol{\eta}_\mathbf{d}(t) - \boldsymbol{\eta}(t)$, $\mathbf{J}^{-1}(\psi)$ is the inverse of the transformation matrix, $\mathbf{K_1}$ is a diagonal matrix that allows quantify control errors, $\mathbf{K_2}$ is a diagonal matrix which is used to delimit the reference velocities, finally $\mathbf{v_c}(t) = \begin{bmatrix} \mu_c & \dot{\psi}_c \end{bmatrix}^\mathsf{T}$ are the reference velocities generated by the controller.

$$\mathbf{v_c}(t) = \mathbf{J}^{-1}(\psi)(\dot{\boldsymbol{\eta}}_\mathbf{d}(t) + \mathbf{K_2} \tanh{(\mathbf{K_2^{-1}K_1}\tilde{\eta}(t))}) \tag{10}$$

Equals Eqs. (10) and (2), we get (11).

$$\mathbf{J}^{-1}(\psi)\dot{\boldsymbol{\eta}}(t) = \mathbf{J}^{-1}(\psi)(\dot{\boldsymbol{\eta}}_\mathbf{d}(t) + \mathbf{K_2} \tanh{(\mathbf{K_2^{-1}K_1}\tilde{\eta}(t))}) \tag{11}$$

This equation is known as a closed-loop representation and considering the velocities error as $\dot{\tilde{\eta}}(t) = \dot{\boldsymbol{\eta}}_\mathbf{d}(t) - \dot{\boldsymbol{\eta}}(t)$, therefore, the final expression would remain expressed as follows (12), this equation is useful to demonstrate the stability of the controller.

$$\dot{\tilde{\eta}}(t) = -\mathbf{K_2} \tanh{(\mathbf{K_2^{-1}K_1}\tilde{\eta}(t))} \tag{12}$$

3.2 Self-tuning Kinematic Controller Gains

In controller based on the inverse kinematics given in Eq. (10), it can be noted that this algorithm presents a matrix of diagonal values $\mathbf{K_1}$, which presents the following structure (13), where K_x is a value proportional to the errors on the x axis, K_y is a value proportional to the errors on the y axis, the disadvantage of this algorithm is that the values must be chosen with care, since they vary depending on the desired trajectories and parametric changes.

$$\mathbf{K_1} = \begin{bmatrix} K_x & 0 \\ 0 & K_y \end{bmatrix} \tag{13}$$

Therefore, a kinematic self-tuning algorithm is based on optimization according to the gradient descent method [6]. The mathematical structure that will be used to adapt the gains will be subject to optimization problems and the equation is given as,

$$\mathbf{G} = \frac{1}{2}\tilde{\eta}^{\mathsf{T}}\, \mathbf{R}\, \tilde{\eta} \tag{14}$$

where $\mathbf{G}$ is the cost function and $\mathbf{R}$ is positive definite diagonal matrix that will weigh the position errors. To find the gains that minimize the cost function, the chain rule method will be used, considering that, for this method to be applied, a vector $\boldsymbol{\kappa} = \begin{bmatrix} K_x & K_y \end{bmatrix}$ is generated, therefore the partial derivative with respect to $\boldsymbol{\kappa}$ is obtained (15).

$$\frac{\partial \mathbf{G}}{\partial \boldsymbol{\kappa}} = \tilde{\eta}^{\mathsf{T}}\, \mathbf{R}\, \frac{\partial \tilde{\eta}}{\partial \boldsymbol{\kappa}} \tag{15}$$

$$\frac{\partial \tilde{\eta}}{\partial \boldsymbol{\kappa}} = -\frac{\partial \eta}{\partial \mathbf{v_c}} \frac{\partial \mathbf{v_c}}{\partial \boldsymbol{\kappa}} \tag{16}$$

The first derivative in the above Eq. (16) can be defined as the Jacobian matrix with the system velocity inputs, which can be calculated using direct neural network model and back-propagation algorithm [9].

The above deduced that the controller's profits change and adapt by making the cost function tend to zero, therefore according to the algorithm of the descent of the gradient, which is shown below represented in (17)

$$\Delta \boldsymbol{\kappa} = -\boldsymbol{\alpha}\, \frac{\partial \mathbf{G}}{\partial \boldsymbol{\kappa}} = \boldsymbol{\alpha}\, \tilde{\eta}^{\mathsf{T}}\, \mathbf{R}\, \frac{\partial \eta}{\partial \mathbf{v_c}} \frac{\partial \mathbf{v_c}}{\partial \boldsymbol{\kappa}} \tag{17}$$

where $\boldsymbol{\alpha}$ is a diagonal matrix with the learning rates of the gradient descent algorithm.

3.3 Artificial Neural Network Direct Model

The artificial neural network that is used to approximate the robot platform is shown in Fig. 4. The algorithm used to obtain is back-propagation method, chosen for its ability to adapt to changing environments [7].

For the implementation of the back-propagation algorithm, it is considered that neural network has $\mathbf{L}$ layers in the internal structure, where $\mathbf{Z_0}$ is the input vector conformed by $\mathbf{Z_0} = \begin{bmatrix} \mu_c & \dot{\psi}_c \end{bmatrix}^{\mathsf{T}}$, $\mathbf{Z_L} = \begin{bmatrix} \bar{x} & \bar{y} \end{bmatrix}^{\mathsf{T}}$ is the output vector and $\boldsymbol{\eta} = \begin{bmatrix} x & y \end{bmatrix}^{\mathsf{T}}$ is

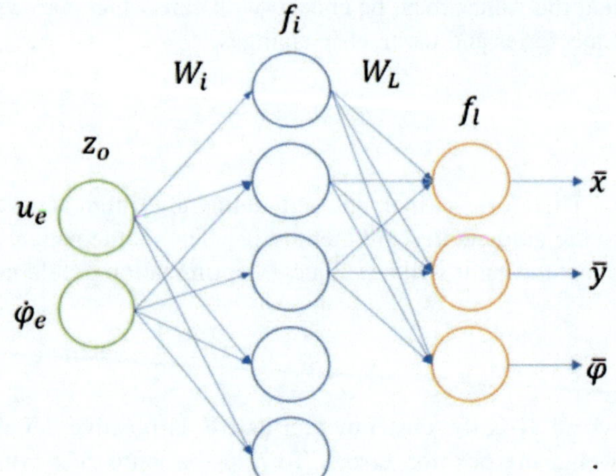

Fig. 4 Structure of the neural network

the vector of the truth that the neural network tries to estimate. The weight matrices of the neural network are $\mathbf{W_1}, \mathbf{W_2}, \ldots\mathbf{W_L}$, and the non-linear activation's functions are $f_1, f_2, \ldots f_L$.

With these considerations, forward propagation in the general form results in (18), the difference between the desired estimation and the values of the neural network is defined in (19) known as the error vector.

$$\mathbf{Z_i} = f_i(\mathbf{W_i Z_{i-1}}) \tag{18}$$

$$\mathbf{E} = \left\|\mathbf{Z_L} - \eta\right\|^2 \tag{19}$$

The back-propagation of the error vector in the general form is (20) and (21). Finally, adaptive weights of the neural network are (22) and (23).

$$\delta_L = (\mathbf{Z_L} - \eta) \otimes f_L^{'}(\mathbf{W_L Z_{L-1}}) \tag{20}$$

$$\delta_i = \mathbf{W}_{i+1}^{\mathsf{T}} \delta_{i+1} \otimes f_i^{'}(\mathbf{W_i Z_{i-1}}) \tag{21}$$

$$\frac{\partial \boldsymbol{E}}{\partial \mathbf{W_i}} = \delta_i \mathbf{Z}_{\mathbf{i-1}}^{\mathsf{T}} \tag{22}$$

$$\mathbf{W_i} = \mathbf{W_i} - \alpha \otimes \frac{\partial \boldsymbol{E}}{\partial \mathbf{W_i}} \tag{23}$$

The Jacobian matrix that can be found from the above neural network is (24), where **I** is a identity matrix.

$$\frac{\partial \bar{\eta}}{\partial \mathbf{v_c}} = \mathbf{W}_{i+1}\,\mathbf{I} \cdot (\mathbf{I}^{\mathsf{T}} \otimes f_i^{'}(\mathbf{W_i Z_{i-1}}))\mathbf{W}_i \tag{24}$$

3.4 Dynamic Compensation

The differential mobile robot presents dynamic effects, so velocities errors are generated $\tilde{\mathbf{v}} = \mathbf{v_c} - \mathbf{v}$, which is the difference between the velocities of the kinematics controller and the real velocities of the robot. To reduce this effect, a dynamic compensation is proposed, which is performed using the dynamic model of the robot.

The control law to be used is indicated in (25), where $\mathbf{K_3}$ it is a constant diagonal matrix proportional to the velocities errors, $\mathbf{K_4}$ it is a diagonal matrix that delimits the velocities sent to the mobile robot.

$$\mathbf{v_{ref}} = \bar{\mathbf{M}}(\mathbf{q})(\dot{\mathbf{v}}_c + \mathbf{K_4}\tanh(\mathbf{K_4^{-1}K_3}\tilde{\mathbf{v}})) + \bar{\mathbf{C}}(\mathbf{q}, \dot{\mathbf{q}})\mathbf{v_c} \tag{25}$$

Grouping (25), (9) and considering the acceleration errors $\dot{\tilde{\mathbf{v}}}(t) = \dot{\mathbf{v}}_c(t) - \dot{\mathbf{v}}(t)$, the Eq. (26) is deduced.

$$\dot{\tilde{\mathbf{v}}} = -\mathbf{K_4}\tanh(\mathbf{K_4^{-1}K_3}\tilde{\mathbf{v}(t)}) \tag{26}$$

3.5 Stability Analysis of Kinematic Controller

For the stability analysis, the Lyapunov method is used, starting with the candidate function of quadratic position errors (27) and applying the temporal derivative is obtained (28).

$$\mathbf{V}(\tilde{\eta}) = \frac{1}{2}\tilde{\eta}(t)^{\mathsf{T}}\tilde{\eta}(t) \tag{27}$$

$$\dot{\mathbf{V}}(\tilde{\eta}) = \tilde{\eta}(t)^{\mathsf{T}}\dot{\tilde{\eta}}(t) \tag{28}$$

By means of (12) and (28) is determined.

$$\dot{\mathbf{V}}(\tilde{\eta}) = -\tilde{\eta}(t)^{\mathsf{T}}\mathbf{K_2}\tanh(\mathbf{K_2^{-1}K_1}\tilde{\eta}(t)) \tag{29}$$

Guaranteeing the stability of the kinematic controller if $\mathbf{K_1} > 0$ and $\mathbf{K_2} > 0$, so that $\tilde{\eta} \to 0$ when $t \to \infty$.

3.6 Stability Analysis of Dynamic Compensation

For the stability analysis, the Lyapunov method is used (30), starting with the candidate function of quadratic velocities errors and applying the temporal derivative is obtained (31).

$$\mathbf{V}(\tilde{\mathbf{v}}) = \frac{1}{2}\tilde{\mathbf{v}}(t)^{\mathsf{T}}\tilde{\mathbf{v}}(t) \tag{30}$$

$$\dot{\mathbf{V}}(\tilde{\mathbf{v}}) = \tilde{\mathbf{v}}(t)^{\mathsf{T}}\dot{\tilde{\mathbf{v}}}(t) \tag{31}$$

By means (31) and (26), (32) is obtained.

$$\dot{\mathbf{V}}(\tilde{\mathbf{v}}) = -\tilde{\mathbf{v}}^{\mathsf{T}}(t)\mathbf{K_4}\tanh\left(\mathbf{K_4^{-1}}\mathbf{K_3}\tilde{\mathbf{v}}(t)\right) \tag{32}$$

Guaranteeing the stability of the dynamic compensation if $\mathbf{K_3} > 0$ and $\mathbf{K_4} > 0$, so that $\tilde{\mathbf{v}} \to 0$ when $t \to \infty$.

4 Results

4.1 Simulation Results

The validation of the proposed controller is carried out through the Webots simulation software [13]. The modeling of the structures and the simulation scenario is generated in the Sketch CAD software, then imported to the Webots graphics engine. The simulation scenario has a circular area with a radius of 3 meters, as shown in Fig. 5.

The collision system, the force of gravity and the coefficient of friction between contact surfaces are included in the physics engine of the simulator. The shadows and the simulation are configured in the rendering at every instant of time. The virtual stage configuration in the scene tree is shown in Table 1.

The Webots simulation software includes CAD models of different types of robots, among which is the Turtlebot3 mobile robot (Fig. 6). The simulated robot is characterized with the same configuration as a real robot.

The reading of the sensors and the actions of the different states of speed, acceleration, among others, are possible due to the libraries that can be programmed in a controller within the Webots simulation software. The Turtlebot3 mobile robot integrates different sensors, such as a reference GPS and a gyroscope that allow knowing the location and orientation respectively. The global and local reference systems are configured with the direction of their axes according to the law of the right hand.

Following the scheme shown in Fig. 7, the connection of the robot is established through ROS nodes, which allow direct communication between the data of the robot states and the mathematical software that implements the control algorithms. A full

Fig. 5 3D simulation scenario. **a** Isometric view. **b** Aerial view. **c** Turtlebot3 mobile robot

Table 1 Webots world configuration

Field scenario	Values
Gravity (x, y, z)	$(0, -9.81, 0)$
Constraint force mixing	0.00005
Error reduction parameter	0.2
Basic time step	64 [ms]
Frames per second	60
Physics disable time	1 [s]
Physics disable linear threshold	0.01 [m/s]
Physics disable angular threshold	0.01 [rad/s]
North direction	$(1, 0, 0)$
Gps reference	$(0, 0, 0)$

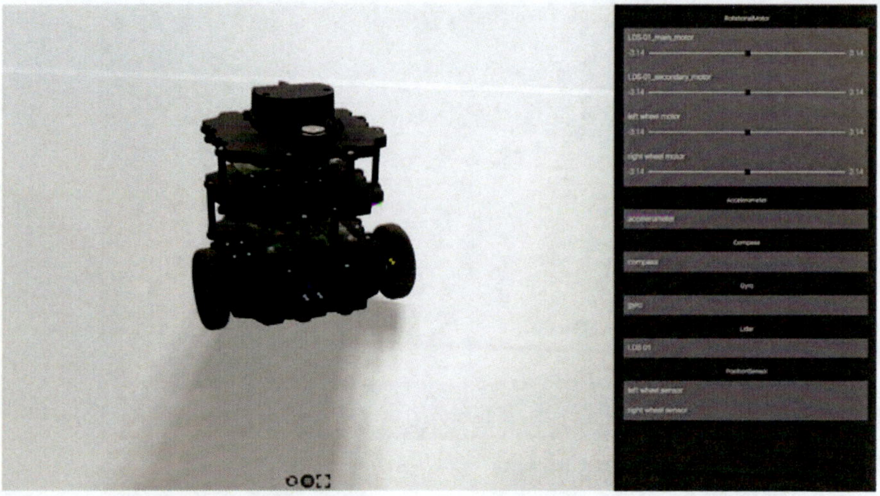

Fig. 6 Turtlebot3 mobile robot Webots

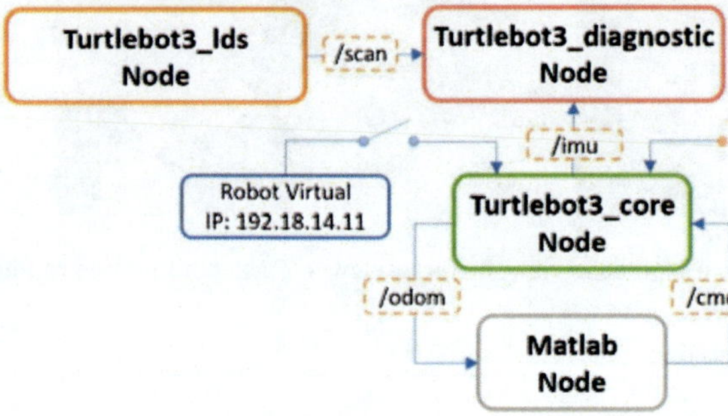

Fig. 7 Communication structure

Table 2 Desired trajectory and initial configuration for simulation

Parameters	Values	Parameters	Values
x_d	$0.5\cos(0.3t)$ [m]	x_o	0.1 [m]
y_d	$0.5\sin(0.3t)$ [m]	y_o	0.0 [m]

simulation and hardware in the Loop scheme is established for the evaluation of the implemented controller.

It consists of following a desired trajectory while analyzing the behavior of Self-Tuning Kinematic controllers under the disturbances, slippage and white noise generated by Webots. The desired trajectory and the initial conditions for the experiment are presented in Table 2.

Table 3 Initial values for controllers in simulation

Parameters	Values	Parameters	Values
$\mathbf{K_1}$	$diag[0.5\ 0.5]$	$\mathbf{K_2}$	$diag[0.2\ 0.2]$
$\mathbf{K_3}$	$diag[1\ 1]$	$\mathbf{K_4}$	$diag[1\ 1]$
$\mathbf{R}$	$diag[1\ 1]$	α	$diag[0.1\ 0.1]$

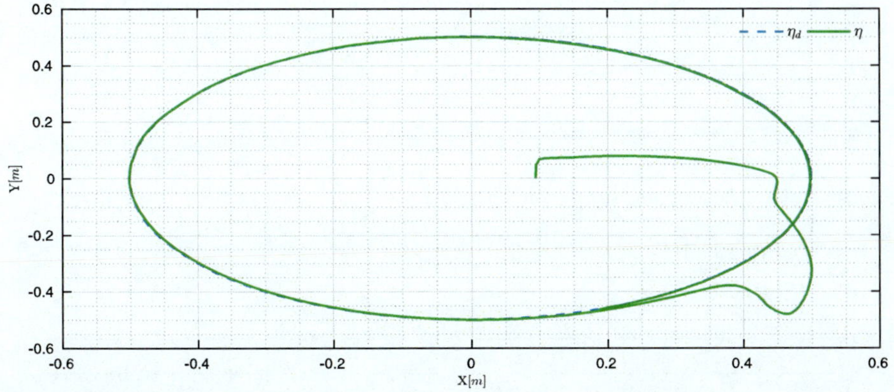

Fig. 8 Movement of the Turtlebot3 in simulation

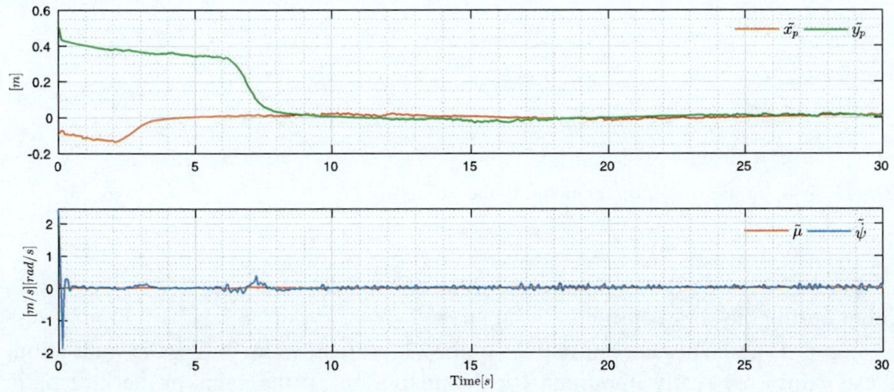

Fig. 9 Control errors in simulation

The gains for the Simulation are defined in Table 3, considering that an initial value of $\mathbf{K_1}$ is established, and $\mathbf{K_2}$ was established under consideration of the limitations of real control velocities.

Figure 8 indicates the behavior of the Turtlebot3 robot in simulator Webots, it can be visualized that the point of interest tends to the established trajectory in Table 2. Figure 9 shows the evolution of the control errors, and it is appreciated that it tends to zero despite the white noise that is introduced into the simulator.

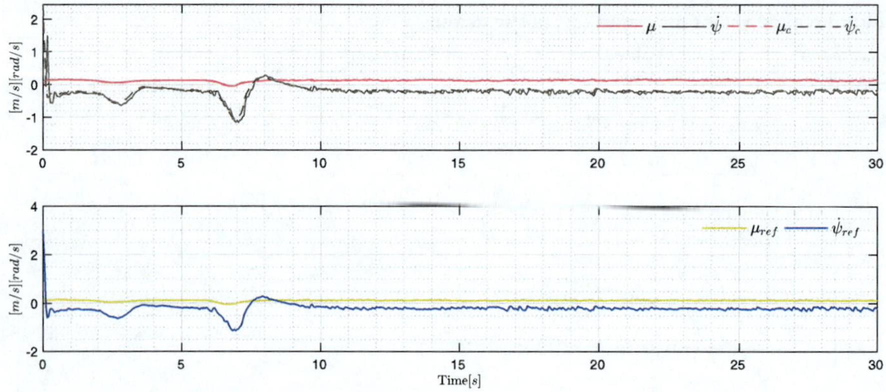

Fig. 10 Control values of the simulation

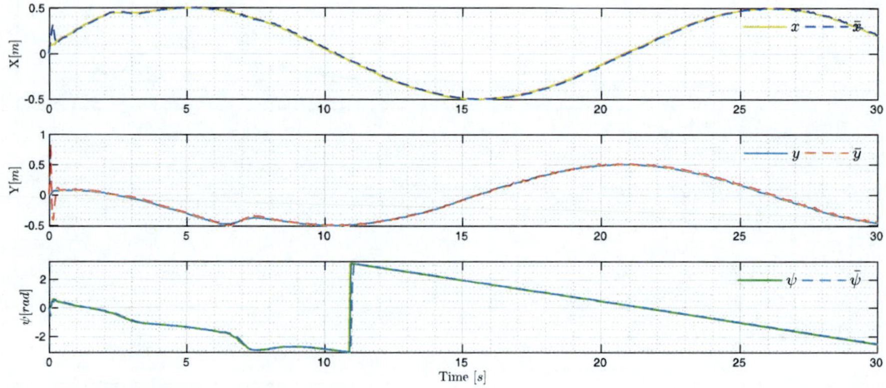

Fig. 11 Estimations of NN direct model of the simulation

Figure 10 shows the kinematic and dynamic control values applied to the mobile robot during the simulation.

Figure 11 indicates the estimations of the states in the robot during the simulation, these values are really important for the adaptation of the gains in the kinematic controller. Figure 12 shows the evolution of the kinematic controller gains.

4.2 Experimental Results

The experimental test was implemented on a TurtleBot3 robot as shown in Fig. 13, which admits linear angular velocities as input reference signals, the controllers were carried out in Matlab software, and the communication was done through robotic operating system ROS, and the same communication structure is shown in Fig. 7.

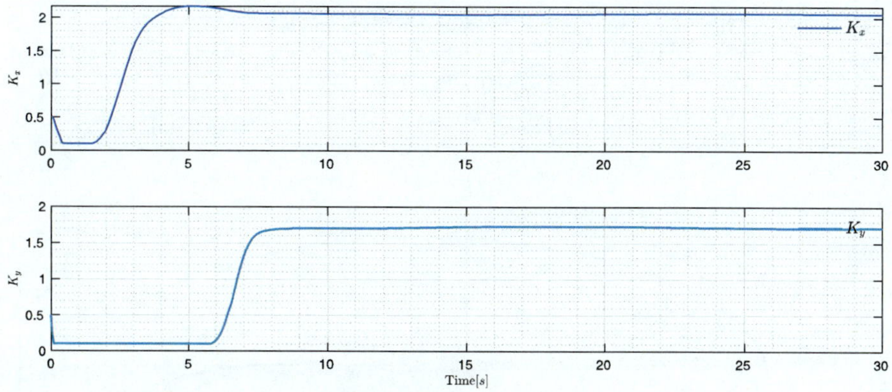

Fig. 12 Adaptive gain evolution of the simulation

Fig. 13 Real Turtlebot3 mobile robot

The desired trajectory and the initial conditions for the experiment are presented in Table 4. The gains for the experiment are the same as the simulation, defined in Table 3.

Figure 14 indicates the behavior of the Turtlebot3 robot in a real environment, it is clear to see that the evolution of the point of interest is tracking the desired trajectory.

Table 4 Desired trajectory and initial configuration for experiment

Parameters	Values	Parameters	Values
x_d	$0.4 \cos(0.3t)$ [m]	x_o	0.1 [m]
y_d	$0.4 \sin(0.3t)$ [m]	y_o	0.0 [m]

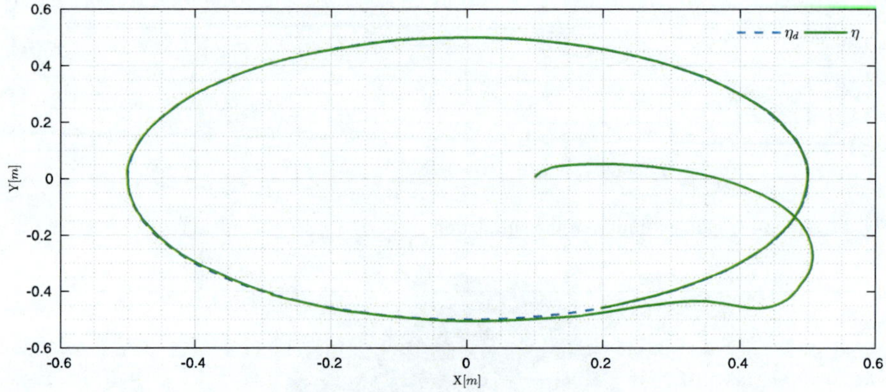

Fig. 14 Movement of the Turtlebot3 in experiment

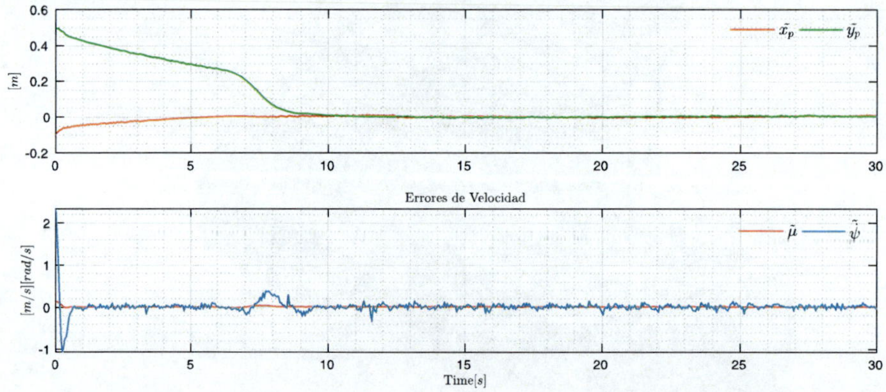

Fig. 15 Control errors in experiment

Figure 15 shows the evolution of the control errors and it is appreciate tends to zero.

The control errors of the dynamic compensation also tend to zero because the controller maintains the values generated by a kinematic controller in the mobile robot.

Figure 16 shows the kinematic control values applied to the mobile robot during the experiment. It is clear to see that the dynamic compensation generates the appro-

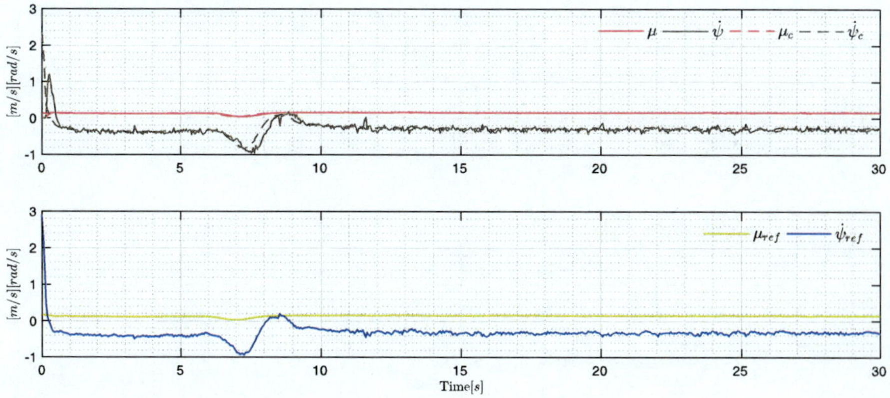

Fig. 16 Control values of the experiment

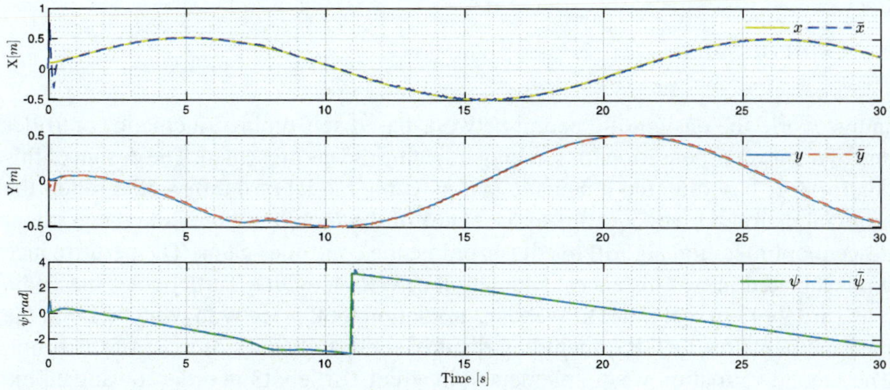

Fig. 17 Estimations of NN direct model of the experiment

priate control values for compensating the dynamics in the real robot, guaranteeing the convergence to zero of the control errors.

Figure 17 indicates that the NN Direct Model makes a correct estimate of the positions and orientations of the mobile robot, which is important when adjusting the controller gains.

Figure 18 shows the evolution of the kinematic controller gains, which adapt considering the desired trajectory and the estimations of the states of the robot in a real environment.

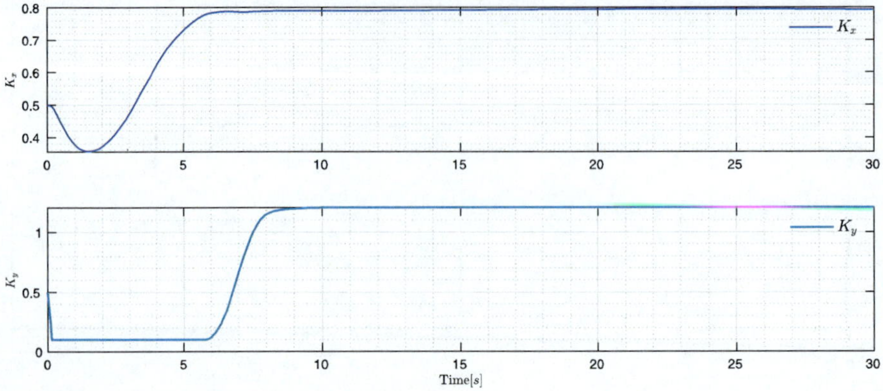

Fig. 18 Adaptive gain evolution of the experiment

5 Conclusions

In this work, the design of a neural network-based self-tuning kinematic controller and dynamic compensation for tracking trajectories was presented. The design of this controller is based on the kinematic and dynamic model of a differential robot, the artificial neural networks estimate the states of the mobile robot while the gradient descent optimization algorithm adjusts online the controller gains. The performance was evaluated in two instances: simulation in Webots with a Turtlebot3 robot and a real Turtlebot3 in real time. In Webots, simulation took place with white noise in the measurements of the GPS, including disturbances of changes of mass of the robot. The second validation was implemented in a real Turtlebot3 in order to validate the controller in a real-time experiment; both experiments show a good performance of the proposed controller.

Acknowledgements The authors would like to thank Proyecto de Investigacion: Análisis, diseño e implementación de algoritmos de control inteligente en controladores con una red de sensores IoT en vehículos para mejorar la seguridad vial; Escuela Superior Politécnica de Chimborazo ESPOCH and the Research Group GITEA, for the support for the development of this work.

References

1. Andaluz GM, Andaluz VH, Terán HC, Arteaga O, Chicaiza FA, Varela J, Ortiz JS, Pérez F, Rivas D, Sánchez JS, Canseco P (2016) Modeling dynamic of the human-wheelchair system applied to NMPC. In: Lecture notes in computer science (including subseries lecture notes in artificial intelligence and lecture notes in bioinformatics). LNCS, vol 9835. Springer, pp 179–190. https://doi.org/10.1007/978-3-319-43518-3_18

2. Andaluz V, Roberti F, Carelli R (2010) Robust control with redundancy resolution and dynamic compensation for mobile manipulators. In: Proceedings of the IEEE international conference on industrial technology, pp 1469–1474. https://doi.org/10.1109/ICIT.2010.5472488
3. Andaluz VH, Canseco P, Varela Aldas J, Ortiz JS, Pérez MG, Roberti F, Carelli R (2014) Robust control with dynamic compensation for human-wheelchair system. Lecture notes in computer science (including subseries lecture notes in artificial intelligence and lecture notes in bioinformatics) 8917:376–389. https://doi.org/10.1007/978-3-319-13966-1_37
4. Bugeja MK, Fabri SG (2007) Dual adaptive control for trajectory tracking of mobile robots. In: Proceedings - IEEE international conference on robotics and automation, pp 2215–2220. https://doi.org/10.1109/ROBOT.2007.363649
5. De La Cruz C, Carelli R (2008) Dynamic model based formation control and obstacle avoidance of multi-robot systems. Robotica 26(3):345–356. https://doi.org/10.1017/S0263574707004092, https://www.cambridge.org/core/product/identifier/S0263574707004092/type/journal_article
6. Fierro R, Lewis FL (1998) Control of a nonholonomic mobile robot using neural networks. IEEE Trans Neural Networks 9(4):589–600. https://doi.org/10.1109/72.701173
7. Gu D, Hu H (2002) Neural predictive control for a car-like mobile robot. Robot Auton Syst 39(2):73–86. https://doi.org/10.1016/S0921-8890(02)00172-0
8. Andaluz VH, Leica P, Roberti F, Toibero M, Carelli R (2012) Adaptive coordinated cooperative control of multi-mobile manipulators. In: Frontiers in advanced control systems. InTech (2012). https://doi.org/10.5772/39143
9. Hernández-Alvarado R, García-Valdovinos L, Salgado-Jiménez T, Gómez-Espinosa A, Fonseca-Navarro F (2016) Neural network-based self-tuning PID control for underwater vehicles. Sensors 16(9):1429. https://doi.org/10.3390/s16091429, http://www.mdpi.com/1424-8220/16/9/1429
10. Kha NB, Ahn KK (2006) Position control of shape memory alloy actuators by using self tuning fuzzy PID controller. In: 2006 1ST IEEE conference on industrial electronics and applications, pp 1–5. https://doi.org/10.1109/ICIEA.2006.257198
11. Liu C, Zhou C, Cao W, Li F, Jia P (2020) A novel design and implementation of autonomous robotic car based on ROS in indoor scenario. Robotics 9(1):19. https://doi.org/10.3390/robotics9010019, https://www.mdpi.com/2218-6581/9/1/19
12. Mannan MA, Murata T, Tamura J, Tsuchiya T (2006) A fuzzy-logic-based self-tuning PI controller for high-performance vector controlled induction motor drive. Electric Power Compon Syst 34(4):471–481
13. Michel O (2004) Cyberbotics Ltd. Webots: professional mobile robot simulation. Int J Adv Robot Syst 1(1):5 (2004). https://doi.org/10.5772/5618, http://journals.sagepub.com/doi/10.5772/5618
14. Osusky J, Ciganek J (2018) Trajectory tracking robust control for two wheels robot. In: Proceedings of the 29th international conference on cybernetics and informatics, K and I 2018. Institute of Electrical and Electronics Engineers Inc, pp 1–4. https://doi.org/10.1109/CYBERI.2018.8337559
15. Park C, Kyung JH, Choi TY, Do HM, Kim BI, Lee SH (2012) Design of an industrial dual arm robot manipulator for a Human-Robot hybrid manufacturing. In: 2012 9th international conference on ubiquitous robots and ambient intelligence, URAI 2012, pp 616–618. https://doi.org/10.1109/URAI.2012.6463097
16. Recalde LF, Guevara BS, Cuzco G, Andaluz VH (2020) Optimal control problem of a differential drive robot. In: Lecture notes in computer science (including subseries lecture notes in artificial intelligence and lecture notes in bioinformatics), vol 12144 LNAI. Springer Science and Business Media Deutschland GmbH, pp 75–82. https://doi.org/10.1007/978-3-030-55789-8_7, https://link.springer.com/chapter/10.1007/978-3-030-55789-8_7
17. Rossomando F, Soria C, Patiño H, Carelli R (2011) Model reference adaptive control for mobile robots in trajectory tracking using radial basis function neural networks. Lat Am Appl Res 41:177–182

18. Varela-Aldas J, Andaluz VH, Chicaiza FA (2018) Modelling and control of a mobile manipulator for trajectory tracking. In: Proceedings - 3rd international conference on information systems and computer science, INCISCOS 2018, vol 2018-Decem. Institute of Electrical and Electronics Engineers Inc, pp 69–74. https://doi.org/10.1109/INCISCOS.2018.00018
19. Velagic J, Osmic N, Lacevic B (2010) Design of neural network mobile robot motion controller. In: Ramov B (ed) New trends in technologies, chap. 10. IntechOpen, Rijeka (2010). https://doi.org/10.5772/7584
20. Zea DJ, Guevara BS, Recalde LF, Andaluz VH (2021) Dynamic simulation and kinematic control for autonomous driving in automobile robots. Springer, Cham, pp 205–216. https://doi.org/10.1007/978-3-030-63665-4_16, http://link.springer.com/10.1007/978-3-030-63665-4_16

H_∞ Observer for Road Profile Estimation in an Automotive Semi-active Suspension System Using Two Accelerometers

Thanh-Phong Pham, Olivier Sename, Cao Tho Phan, and Gia Quoc Bao Tran

Abstract This work presents an H_∞ observer design for road profile estimation in the automotive semi-active suspension system. The dynamics of the quarter-car augmented with a nonlinear dynamic model of the semi-active damper are written into the descriptor system with road profile as the system states. Then an H_∞ observer is developed to estimate the road profile using the onboard accelerometers as the observer's input. The objective is to minimize the effect of sensor noises on the estimation error using H_∞ framework. The estimation approach is simulated on the quarter-car model of a scaled testbench of Gipsa-lab. Simulation results show the effectiveness of the proposed method.

Keywords H_∞ observer · Road profile estimation · Descriptor system

1 Introduction

In automotive applications, the road profile is considered one of the main factors influencing the vehicle system's performance. Therefore, the real-time knowledge of road input plays a vital role in the automotive suspension control system (see [13, 14, 16] and references therein). Many suspension control approaches are developed under the standard assumption of real-time access (through measurement or estimation

T.-P. Pham (✉) · C. Tho Phan
The University of Danang—University of Technology and Education, Danang 550000, Vietnam
e-mail: ptphong@ute.udn.vn

C. Tho Phan
e-mail: pctho@ute.udn.vn

O. Sename · G. Q. Bao Tran
University Grenoble Alpes, CNRS, Grenoble INP, Institute of Engineering, GIPSA-lab, 38000 Grenoble, France
e-mail: olivier.sename@gipsa-lab.grenoble-inp.fr

G. Q. Bao Tran
e-mail: gia-quoc-bao.tran@grenoble-inp.org

G. Conte and O. Sename (eds.), *Proceedings of 10th International Conference on Mechatronics and Control Engineering*, Lecture Notes in Mechanical Engineering,
https://doi.org/10.1007/978-981-19-1540-6_12

methods) of road disturbance such as [6, 15], in which, the profile measurements based either on profilometers presented in [1, 9] or on visual inspection in [18]. As an alternative and preventive, the estimation strategies are utilized to estimate the road profile since its direct measurement is more expensive.

Many estimation approaches are introduced to estimate the road profile using onboard sensors. The work in [17] presented an artificial neural network (ANN) based method to estimate the road excitation, using accelerometers. Besides, [20] developed H_∞ observer considering the road profile as an unknown input to estimate the variables of the suspension system in the first step. The second step was using the above-estimated states, the unsprung mass accelerometer data, and the quarter-car model to obtain the road profile. On the other hand, [5, 19] presented the adaptive estimation method based on Youla-Kučera (YK) parameterization for road profile estimation and classified the road roughness according to ISO standard using Fourier transform. Although the presented results are exciting, it is also worth noting that the above method required the sprung mass displacement sensor as an input.

On the other hand, using a state observer to estimate the road profile needs a differential equation for the road profile variable. However, it is not easy to model for any road profile. Some models are today used as in [7, 19] but these are valid for a specific ISO classification only. In order to deal with this changeling, the dynamic system can be modeled as a descriptor system. Then we can use the robust observer for the descriptor system presented in [3, 4, 8, 10] to estimate the road profile. The method given in this work is an H_∞ observer to estimate the road profile using two accelerometers of the quarter-car model. First, the dynamic system needs to represent a descriptor system to avoid using the differential equation of road profile while the nonlinearity in the damper model satisfies a Lipschitz condition. An H_∞ observer is then developed, following the parameterization steps in [4], in which the H_∞ framework is used to minimize the effect of sensor noises on the estimation error and the nonlinearity in the dynamic model is bounded via Lipschitz condition. The main contributions of this work are as follows:

- We do not need any equation describing the road profile since the quarter-car model with a dynamical nonlinear damper model is written into a nonlinear descriptor system. The H_∞ observer for the nonlinear descriptor system is then developed to estimate the road profile, minimizing the sensor noise effect on the estimation errors.
- The proposed observer has been simulated on a scaled-vehicle test bench model, which model evaluated by the experimental tests.

The rest of this paper is as follows. Section 2 presents the dynamic model of the quarter car suspension system and the nonlinear descriptor reformulation. Section 3 develops the H_∞ observer design. In Sect. 4, the method is analyzed in the frequency domain. Some simulation results in the time domain are presented in Sect. 5. Finally, Sect. 6 gives some concluding remarks.

2 Semi-active Suspension Modeling and Quarter-Car System Description

2.1 Quarter-Car System Description

This section presents the quarter-car model with the semi-active ER suspension system. The well-known model consists of the sprung mass (m_s), the unsprung mass (m_{us}), the suspension components located between (m_s) and (m_{us}) and the tire, which is modeled as a spring with stiffness k_t. From Newton's second law of motion, the system dynamics around the equilibrium are given as:

$$\begin{cases} m_s\ddot{z}_s &= -F_s - F_d \\ m_{us}\ddot{z}_{us} &= F_s + F_d - F_t \end{cases} \tag{1}$$

where $F_s = k_s(z_s - z_{us})$ is the spring force; $F_t = k_t(z_{us} - z_r)$ is the tire force; the damper force F_d is given as follows (see [11, 12]):

$$\begin{cases} F_d &= k_0(z_s - z_{us}) + c_0(\dot{z}_s - \dot{z}_{us}) + F_{er} \\ \dot{F}_{er} &= -\frac{1}{\tau}F_{er} + \frac{f_c}{\tau} \cdot u \cdot tanh(k_1(z_s - z_{us}) + c_1(\dot{z}_s - \dot{z}_{us})) \end{cases} \tag{2}$$

where $c_0, c_1, k_0, k_1, f_c, \tau$ are constant parameters; z_s and z_{us} are the displacements of the sprung and unsprung masses, respectively. z_r is the road displacement input.

All of the system's parameters are shown in Table 1.

Table 1 Parameter values of the quarter-car model equipped with an ER damper

Parameter	Description	Value	Unit
m_s	Sprung mass	2.27	kg
m_{us}	Unsprung mass	0.25	kg
k_s	Spring stiffness	1396	N/m
k_t	Tire stiffness	12270	N/m
k_0	Passive damper stiffness coefficient	170.4	N/m
c_0	Viscous damping coefficient	68.83	N.s/m
k_1	Hysteresis coefficient due to displacement	218.16	N.s/m
c_1	Hysteresis coefficient due to velocity	21	N.s/m
f_c	Dynamic yield force of ER fluid	28.07	N
τ	Time constant	43	ms

2.2 Descriptor System Modeling

By selecting the system states as $x = [x_1, x_2, x_3, x_4, x_5, x_6]^T = [z_s, \dot{z}_s, z_{us}, \dot{z}_{us}, F_{er}, z_r]^T \in \mathbb{R}^{n_x}$, the measured variables $y = [\ddot{z}_s, \ddot{z}_{us}]^T \in \mathbb{R}^{n_y}$ (notice that in our application $n_x = 6$ and $n_y = 2$), the system dynamics can be written in the following descriptor system form:

$$\begin{cases} E\dot{x} &= Ax + B\Phi(Ex) \cdot u \\ y &= Cx + D\omega \end{cases} \tag{3}$$

where ω is the sensor noises. The nonlinear function $\Phi(Ex) = tanh(k_1x_1 + c_1(x_2 - x_4)) = tanh(\Gamma_e x)$ with $\Gamma_e = \begin{bmatrix} k_1, \ c_1, \ -k_1, \ -c_1, \ 0 \ 0 \end{bmatrix}$. Notice that the nonlinear function $\Phi(Ex)$ satisfies the Lipschitz condition in x

$$\|\Phi(Ex) - \Phi(E\hat{x})\| \leqslant \|\Gamma_e(x - \hat{x})\|, \forall x, \hat{x} \tag{4}$$

where

$$E = \begin{bmatrix} 1\,0\,0\,0\,0\,0 \\ 0\,1\,0\,0\,0\,0 \\ 0\,0\,1\,0\,0\,0 \\ 0\,0\,0\,1\,0\,0 \\ 0\,0\,0\,0\,1\,0 \end{bmatrix}, A = \begin{bmatrix} 0 & 1 & 0 & 0 & 0 & 0 \\ -\frac{(k_s+k_0)}{m_s} & -\frac{c_0}{m_s} & \frac{(k_s+k_0)}{m_s} & \frac{c_0}{m_s} & -\frac{1}{m_s} & 0 \\ 0 & 0 & 0 & 1 & 0 & 0 \\ \frac{(k_s+k_0)}{m_{us}} & \frac{c_0}{m_{us}} & -\frac{(k_s+k_0+k_t)}{m_{us}} & -\frac{c_0}{m_{us}} & \frac{1}{m_{us}} & \frac{k_t}{m_{us}} \\ 0 & 0 & 0 & 0 & -\frac{1}{\tau} & 0 \end{bmatrix}$$

$$C = \begin{bmatrix} -\frac{(k_s+k_0)}{m_s} & -\frac{c_0}{m_s} & \frac{(k_s+k_0)}{m_s} & \frac{c_0}{m_s} & -\frac{1}{m_s} & 0 \\ \frac{(k_s+k_0)}{m_{us}} & \frac{c_0}{m_{us}} & -\frac{(k_s+k_0+k_t)}{m_{us}} & -\frac{c_0}{m_{us}} & \frac{1}{m_{us}} & \frac{k_t}{m_{us}} \end{bmatrix}, B = \begin{bmatrix} 0 \\ 0 \\ 0 \\ 0 \\ \frac{f_c}{\tau} \end{bmatrix}, D = \begin{bmatrix} 0.01 \\ 0.01 \end{bmatrix}$$

3 H_∞ Observer Design

The reduced-order H_∞ observer for the quarter-car system (3) is chosen as:

$$\begin{cases} \dot{z} = Nz + Jy + H\Phi(E\hat{x}) \cdot u \\ \hat{x} = Rz + Sy \end{cases} \tag{5}$$

where $z \in \mathbb{R}^{n_x - n_y}$ is the state variable of the observer, here $z \in \mathbb{R}^4$, $\hat{x}$ is the estimated state of x. The observer matrices N, L, H, J, S appropriate dimensions have to be designed.

The dynamic error is given as follows:

$$\epsilon = z - TEx, \tag{6}$$

where the matrix T is an arbitrary matrix.

Differentiating (6) with respect to time and using (3) and (5), leads to:

$$\begin{cases} \dot{\epsilon} & = \dot{z} - TE\dot{x} \\ & = N\epsilon + (NTE - TA + JC)x + JD\omega + (H - TB) \cdot \Phi(E\hat{x})u \\ & -TB \cdot (\Phi(Ex) - \Phi(\hat{E}x))u \\ \hat{x} & = R\epsilon + (RTE + SC)x + SD\omega. \end{cases} \tag{7}$$

It is obvious that if the following decoupling conditions are satisfied:

$$NTE - TA + JC = 0 \tag{8}$$

$$H - TB = 0 \tag{9}$$

$$RTE + SC = I \tag{10}$$

the system (7) becomes

$$\begin{cases} \dot{\epsilon} & = N\epsilon - TB\Delta\Phi + JD\omega \\ e & = R\epsilon + SD\omega \end{cases} \tag{11}$$

where $e = \hat{x} - x$ is the state estimation error and $\Delta\Phi = \Phi(Ex) - \Phi(E\hat{x}) \cdot u$.

In our application, u is the duty cycle of PWM signal. Therefore, it is worth noting that the maximum value of u is 1. Therefore, the term $\Delta\Phi$ is bounded by $0 \leq \Delta\Phi \leq \Phi(Ex) - \Phi(E\hat{x})$.

The problem of the $\mathcal{H}_\infty$ observer design reduces to determine the observer matrices N, J, H, R, S such that all conditions (8)–(10) are satisfied and the effect of measurement noise ω on the state estimation error e is minimized while $\Delta\Phi$ is bounded.

3.1 Parameterization of the Observer Matrices

In order to determine the observer matrices N, J, H, R, S of the proposed observer satisfying all the conditions equalities (8)–(10), the parameterization is made by using the general solution of (8) and (10). The parameterization steps in here are similar with [4].

Firstly, from Eqs. (8) and (10), one obtained

$$\begin{pmatrix} N & J \\ R & S \end{pmatrix} \begin{pmatrix} TE \\ C \end{pmatrix} = \begin{pmatrix} TA \\ I \end{pmatrix}. \tag{12}$$

Equation (12) is solvable if and only if

$$rank\begin{pmatrix} TE \\ C \\ TA \\ I \end{pmatrix} = rank\begin{pmatrix} TE \\ C \end{pmatrix} = n_x. \tag{13}$$

Next, let matrix $M \in \mathbb{R}^{n_x \times n_x}$ be an arbitrary matrix of full row rank such that:

$$rank\begin{pmatrix} M \\ C \end{pmatrix} = rank\begin{pmatrix} TE \\ C \end{pmatrix} = n_x. \tag{14}$$

Then there always exists a parameter matrix K such that:

$$\begin{pmatrix} TE \\ C \end{pmatrix} = \begin{pmatrix} I & -K \\ 0 & I \end{pmatrix}\begin{pmatrix} M \\ C \end{pmatrix} \Leftrightarrow TE = M - KC \tag{15}$$

$$\Rightarrow \begin{pmatrix} T & K \end{pmatrix}\begin{pmatrix} E \\ C \end{pmatrix} = M. \tag{16}$$

A solution for (14) is given by

$$\begin{pmatrix} T & K \end{pmatrix} = M\Sigma^{+} \tag{17}$$

where $\Sigma = \begin{pmatrix} E & C \end{pmatrix}$, Σ^{+} is any general inverse of matrix Σ satisfying $\Sigma\Sigma^{+}\Sigma = \Sigma$. This is equivalent to:

$$T = M\Sigma^{+}\begin{pmatrix} I \\ 0 \end{pmatrix}, \quad K = M\Sigma^{+}\begin{pmatrix} 0 \\ I \end{pmatrix}. \tag{18}$$

Besides, the solution set of (12) is given by

$$\begin{pmatrix} N & J \\ R & S \end{pmatrix} = \begin{pmatrix} TA \\ I \end{pmatrix}\begin{pmatrix} TE \\ C \end{pmatrix}^{+} + \begin{pmatrix} Z_1 \\ Z_2 \end{pmatrix}\left(I - \begin{pmatrix} TE \\ C \end{pmatrix}\begin{pmatrix} TE \\ C \end{pmatrix}^{+}\right) \tag{19}$$

where $\begin{pmatrix} Z_1 \\ Z_2 \end{pmatrix}$ is a free matrix of appropriate dimension. This is equivalent to:

$$N = TA\alpha_1 + Z_1\beta_1 \tag{20}$$
$$J = TA\alpha_2 + Z_1\beta_2 \tag{21}$$
$$R = \alpha_1 + Z_2\beta_1 \tag{22}$$
$$S = \alpha_2 + Z_2\beta_2 \tag{23}$$

where $\alpha_1 = \begin{pmatrix} TE \\ C \end{pmatrix}^+ \begin{pmatrix} I \\ 0 \end{pmatrix}, \alpha_2 = \begin{pmatrix} TE \\ C \end{pmatrix}^+ \begin{pmatrix} 0 \\ I \end{pmatrix}, \beta_1 = \left(I - \begin{pmatrix} TE \\ C \end{pmatrix}\begin{pmatrix} TE \\ C \end{pmatrix}^+\right)\begin{pmatrix} I \\ 0 \end{pmatrix}$,
$\beta_2 = \left(I - \begin{pmatrix} TE \\ C \end{pmatrix}\begin{pmatrix} TE \\ C \end{pmatrix}^+\right)\begin{pmatrix} 0 \\ I \end{pmatrix}$.

Remark: If the matrices N, J, R, S, H can be chosen according to (20), (21), (22), (23) and (9), respectively, then, all conditions (8)–(10) are fulfilled.

From the results of above parameterization, for simplicity, the matrices of system (11) can be rewritten as follows:

$$\mathbb{A}_1 = N = A_{11} + Z_1 A_{12} \tag{24}$$
$$\mathbb{B}_1 = JD = B_{11} + Z_1 B_{12} \tag{25}$$
$$\mathbb{W}_1 = -TB \tag{26}$$
$$\mathbb{C}_1 = R = C_{11} + Z_2 C_{12} \tag{27}$$
$$\mathbb{D}_1 = SD = D_{11} + Z_2 D_{12} \tag{28}$$

where $A_{11} = TA\alpha_1$, $A_{12} = \beta_1$, $B_{11} = TA\alpha_2 D$, $B_{12} = \beta_2 D$, $C_{11} = \alpha_1$, $C_{12} = \beta_1$, $D_{11} = \alpha_2 D$, $D_{12} = \beta_2 D$.

Notice that all the matrices $A_{11}, A_{12}, B_{11}, B_{12}, \mathbb{W}_1, C_{11}, C_{12}, D_{11}, D_{12}$ are known.

3.2 $\mathcal{H}_\infty$ Design

Using (24)–(28), the estimation error dynamic system (11) is rewritten as:

$$\begin{cases} \dot{\epsilon} &= \mathbb{A}_1 \epsilon + \mathbb{W}_1 \Delta\Phi + \mathbb{B}_1 \omega \\ e &= \mathbb{C}_1 \epsilon + \mathbb{D}_1 \omega. \end{cases} \tag{29}$$

Assuming the Lipschitz condition (4) for $\Phi(x)$, the $\mathcal{H}_\infty$ observer design problem is to determine the matrix Z_1 and Z_2 such that:

- The system (29) is asymptotically stable for $\omega(t) = 0$
- Minimize γ such that $\|e(t)\|_{\mathcal{L}_2} < \gamma\|\omega(t)\|_{\mathcal{L}_2}$ for $\omega(t) \neq 0$

The following theorem solves the above problem into an LMI framework.

Theorem 1 *Consider the system model (3) and the observer (5). The above design problem is solved if there exist a symmetric positive definite matrix X, matrices Y, Z_2 and positive scalar ϵ_1 minimizing γ such that:*

$$\begin{bmatrix} \Omega & X\mathbb{W}_1 & XB_{11}+YB_{12} & C_{11}^T+C_{12}^TZ_2^T & C_{11}^T\Gamma_e^T+C_{12}^TZ_2^T\Gamma_e^T \\ \mathbb{W}_1^TX & -\epsilon_l I & 0 & 0 & 0 \\ B_{11}^TX+B_{12}^TY^T & 0 & -\gamma^2 I & D_{11}^T+D_{12}^TZ_2^T & D_{11}^T\Gamma_e^T+D_{12}^TZ_2^T\Gamma_e^T \\ C_{11}+Z_2C_{12} & 0 & D_{11}+Z_2D_{12} & -I & 0 \\ \Gamma_eC_{11}+\Gamma_eZ_2C_{12} & 0 & \Gamma_eD_{11}+\Gamma_eZ_2D_{12} & 0 & -\epsilon_l I \end{bmatrix} < 0 \tag{30}$$

where $\Omega = A_{11}^T X + XA_{11} + A_{12}^T Y^T + YA_{12}$, *the matrix* Z_1 *is then deduced as* $Z_1 = X^{-1}Y$.

Proof Consider the following Lyapunov function candidate

$$V(\cdot) = \epsilon(\cdot)^T X \epsilon(\cdot). \tag{31}$$

Differentiating $V(\cdot)$ along the solution of (29) yields

$$\begin{aligned} \dot{V}(\cdot) &= \dot{\epsilon}^T X\epsilon + \epsilon^T X\dot{\epsilon} = \epsilon^T(\mathbb{A}_1^T X + X\mathbb{A}_1)\epsilon + \Delta\Phi^T \mathbb{W}_1(\rho)^T X\epsilon + \omega^T \mathbb{B}_1^T X\epsilon \\ &+ \epsilon^T X\mathbb{W}_1(\rho)\Delta\Phi + \epsilon^T X\mathbb{B}_1\omega \end{aligned} \tag{32}$$

In order to satisfy the $\mathcal{H}_\infty$ performance objective w.r.t. the $\mathcal{L}_2$ gain disturbance attenuation, we need to consider the following inequality

$$\begin{aligned} &\dot{V} + e^T e - \gamma^2 \omega^T \omega < 0 \\ \Leftrightarrow &\begin{bmatrix} \epsilon^T \\ \Delta\Phi^T \\ \omega^T \end{bmatrix}^T \begin{bmatrix} \mathbb{A}_1^T X + X\mathbb{A}_1 + \mathbb{C}_1^T\mathbb{C}_1 & X\mathbb{W}_1 & X\mathbb{B}_1 + \mathbb{C}_1^T\mathbb{D}_1 \\ \mathbb{W}_1^T X & 0 & 0 \\ \mathbb{B}_1^T X + \mathbb{D}_1^T\mathbb{C}_1 & 0 & \mathbb{D}_1^T\mathbb{D}_1 - \gamma^2 I \end{bmatrix} \begin{bmatrix} \epsilon \\ \Delta\Phi \\ \omega \end{bmatrix} < 0 \end{aligned} \tag{33}$$

Defining $\eta = \begin{bmatrix} \epsilon \\ \Delta\Phi \\ \omega \end{bmatrix}$, one obtains

$$\eta^T M \eta < 0 \tag{34}$$

where $M = \begin{bmatrix} \mathbb{A}_1^T X + X\mathbb{A}_1 + \mathbb{C}_1^T\mathbb{C}_1 & X\mathbb{W}_1 & X\mathbb{B}_1 + \mathbb{C}_1^T\mathbb{D}_1 \\ \mathbb{W}_1^T X & 0 & 0 \\ \mathbb{B}_1^T X + \mathbb{D}_1^T\mathbb{C}_1 & 0 & \mathbb{D}_1^T\mathbb{D}_1 - \gamma^2 I \end{bmatrix}$.

From the Lipschitz condition (4) and u bounded by 1, the following condition is obtained

$$(\Delta\Phi)^T \Delta\Phi \leqslant (\mathbb{C}_1\epsilon + \mathbb{D}_1\omega)^T \Gamma_e^T \Gamma_e (\mathbb{C}_1\epsilon + \mathbb{D}_1\omega) \Leftrightarrow \eta^T Q \eta \leqslant 0 \tag{35}$$

where $Q = \begin{bmatrix} -\mathbb{C}_1^T\Gamma_e^T\Gamma_e\mathbb{C}_1 & 0 & -\mathbb{C}_1^T\Gamma_e^T\Gamma_e\mathbb{D}_1 \\ 0 & I & 0 \\ -\mathbb{D}_1^T\Gamma_e^T\Gamma_e\mathbb{C}_1 & 0 & -\mathbb{D}_1^T\Gamma_e^T\Gamma_e\mathbb{D}_1 \end{bmatrix}$

By applying the $\mathcal{S}$-procedure [2] to the contraint (35) and the condition $\dot{V}(\cdot) + e^T e - \gamma^2 \omega^T \omega < 0$ if there exists a scalar $\epsilon_l > 0$ such that

$$\eta^T (M - \epsilon_l Q)\eta < 0 \tag{36}$$

The condition (36) is equivalent to

$$\begin{bmatrix} \Omega_1 + \mathbb{C}_1^T \mathbb{C}_1 + \epsilon_l \mathbb{C}_1^T \Gamma_e^T \Gamma_e \mathbb{C}_1 & X\mathbb{W}_1 & X\mathbb{B}_1 + \mathbb{C}_1^T \mathbb{D}_1 + \epsilon_l \mathbb{C}_1^T \Gamma_e^T \Gamma_e \mathbb{D}_1 \\ \mathbb{W}_1^T X & -\epsilon_l I & 0 \\ \mathbb{B}_1^T X + \mathbb{D}_1^T \mathbb{C}_1 + \epsilon \mathbb{D}_1^T \Gamma_e^T \Gamma_e \mathbb{C}_1 & 0 & \mathbb{D}_1^T \mathbb{D}_1 + \epsilon \mathbb{D}_1^T \Gamma_e^T \Gamma_e \mathbb{D}_1 - \gamma^2 I \end{bmatrix} < 0 \tag{37}$$

where $\Omega_1 = \mathbb{A}_1^T X + X\mathbb{A}_1$.

By applying Schur complement to (37), one obtains

$$\begin{bmatrix} \Omega_1 & X\mathbb{W}_1 & X\mathbb{B}_1 & \mathbb{C}_1^T & \mathbb{C}_1^T \Gamma_e^T \\ \mathbb{W}_1^T X & -\epsilon_l I & 0 & 0 & 0 \\ \mathbb{B}_1^T X & 0 & -\gamma^2 I & \mathbb{D}_1^T & \mathbb{D}_1^T \Gamma_e^T \\ \mathbb{C}_1 & 0 & \mathbb{D}_1 & -I & 0 \\ \Gamma_e \mathbb{C}_1 & 0 & \Gamma_e \mathbb{D}_1 & 0 & -\epsilon_l I \end{bmatrix} < 0 \tag{38}$$

Substituting (24)–(28) into (38), the following inequality is obtained

$$\begin{bmatrix} \Omega_2 & X\mathbb{W}_1 & X(B_{11} + Z_1 B_{12}) & (C_{11} + Z_2 C_{12})^T & (C_{11} + Z_2 C_{12})^T \Gamma_e^T \\ \mathbb{W}_1^T X & -\epsilon_l I & 0 & 0 & 0 \\ (B_{11} + Z_1 B_{12})^T X & 0 & -\gamma^2 I & (D_{11} + Z_2 D_{12})^T & (D_{11} + Z_2 D_{12})^T \Gamma_e^T \\ C_{11} + Z_2 C_{12} & 0 & D_{11} + Z_2 D_{12} & -I & 0 \\ \Gamma_e (C_{11} + Z_2 C_{12}) & 0 & \Gamma_e (D_{11} + Z_2 D_{12}) & 0 & -\epsilon_l I \end{bmatrix} < 0 \tag{39}$$

where $\Omega_2 = (A_{11} + Z_1 A_{12})^T X + X(A_{11} + Z_1 A_{12})$.
Let define $Y = XZ_1$ and substitute into (39), the LMI (30) is obtained.
If (30) is satisfied, from (35), (36) implies that

$$\dot{V} + e^T e - \gamma^2 \omega^T \omega < 0. \tag{40}$$

Under the zero initial conditions, taking the integration of (40), we obtain

$$\|e(t)\|_{\mathcal{L}_2}^2 < \gamma^2 \|\omega(t)\|_{\mathcal{L}_2}^2. \tag{41}$$

The proof of Theorem 1 is completed. □

The H_∞ observer design steps are summarized as the following algorithm:
Input: The system matrices E A, B, C, D
Output: The observer matrices N, J, H, R, S
Step 1: Check the rank condition,
• If $rank\begin{pmatrix} E \\ C \end{pmatrix} = n_x$, continue step 2.
• If $rank\begin{pmatrix} E \\ C \end{pmatrix} < n_x$, stop.
Step 2: Choose the full rank matrix $M \in \mathbb{R}^{n_x \times n_x}$ according to the condition (14), i.e.

$$rank\begin{pmatrix} M \\ C \end{pmatrix} = n_x$$

.
Step 3: Compute matrices K, T, according to the Eq. (18). Then compute the matrices α_1, α_2, β_1, β_2 as explained in Eqs. (20) and (23).
Step 4: Calculate the matrices A_{11}, A_{12}, B_{11}, B_{12}, $\mathbb{W}_1$, C_{11}, C_{12}, D_{11}, D_{12} according to the Eqs. (24)–(28).
Step 5: Solve the LMI (30) to find the solution X, Y, Z_2 γ, ϵ_l. Then use X and Y to get the matrix Z_1 following Theorem 1.
Step 6: Determine the observer matrices (N, J, H, R, S) by using the matrices Z_1 and Z_2 obtained in steps 5.

4 Synthesis Results and Frequency Domain Analysis

In this section, the proposed method is applied to estimate the road profile in the automotive suspension system.

4.1 Synthesis Results

In the INOVE testbed available at GIPSA-lab, the applied control signal u (duty cycle of PWM signal) is taken values in the range of [0, 1]. By applying the Algorithm 1, we obtain the L_2- induced gain $\gamma = 0.967$, $\epsilon_l = 36$ and the observer matrices N, J, H, R, S.

4.2 Frequency Domain Analysis

The resulting attenuation of the sensor noises on the estimation errors is shown in Fig. 1. According to Fig. 1, we can see that the proposed method highlights the

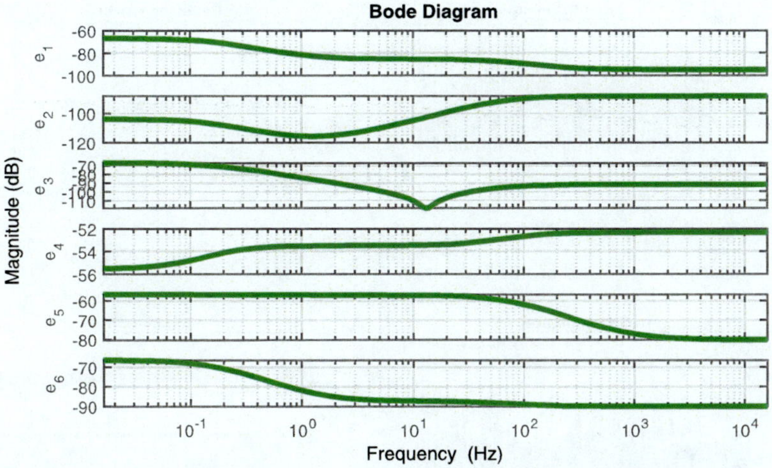

Fig. 1 Transfer $\|e/\omega\|$—Bode diagrams of the proposed observer w.r.t. measurement noise

effectiveness of measurement noise attenuation, which is indicated in the transfer function e_6/ω.

5 Simulation Result

In this section, three simulations performed with the descriptor system (3) are assessed in the time-domain framework. The initial state conditions for the quarter car system (x_0), the proposed observer (z_0) are chosen as the following:

$$x_0 = \left[0,\ 0,\ 0,\ 0,\ 0\right]^T, z_0 = \left[10\ 0.1\ 0.1\ 10\right]^T.$$

In the first scenario, sinusoidal road profile is used. In this simulation, the road profile input is sinusoidal signal with the amplitude at 10^{-3} (m), and the control input u is constant ($u = 0.1$).

The road estimation results of this scenario are presented in Fig. 2. The estimation error is shown on the right side of Fig. 2. The NRMSE of this simulation scenario is shown in Table 2. These results highlight the effectiveness of the proposed observer.

In the second scenario, ISO road profile is used. In this simulation scenario, the ISO road profile (type C) is used and the control input u is obtained from a Skyhook controller.

The simulation results of the second test are shown in Fig. 3. To further describe this accuracy, Table 2 presents the normalized root-mean-square errors, considering the difference between the estimated and measured.

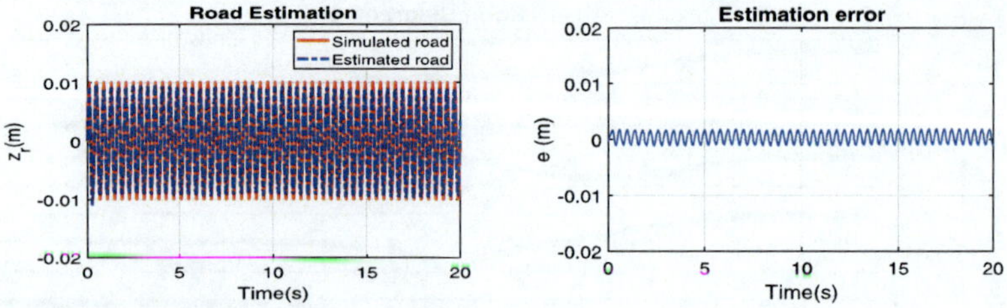

Fig. 2 Scenario 1: (left) road profile estimation, (right) estimation error

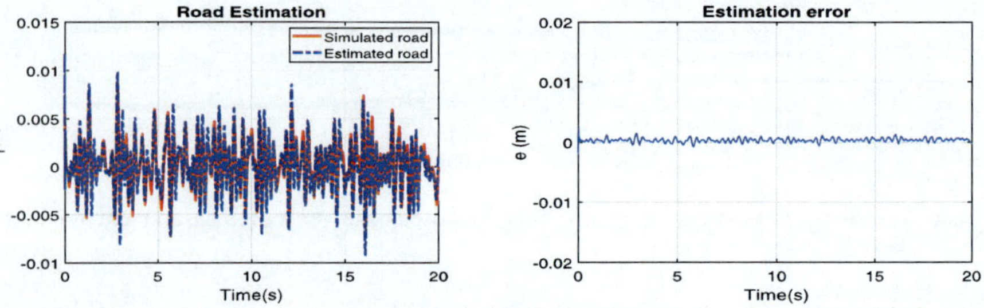

Fig. 3 Scenario 2: (left) road profile estimation, (right) estimation error

Table 2 Normalized Root-Mean-Square Errors (NRMSE) of the simulation scenarios

Scenario	Proposed method
Scenario 1	3.97
Scenario 2	1.67
Scenario 3	5.97

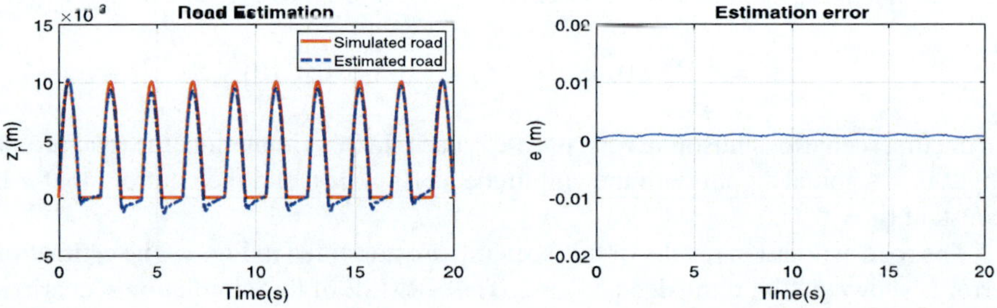

Fig. 4 Scenario 3: (left) road profile estimation, (right) estimation error

In the third scenario, bump road profile is used. In this test, the bump road profile is used, and the control input u is constant ($u = 0.1$).

The estimation results of the third test are shown in Fig. 4. According to these figures, we can see that the proposed observer can estimate the bump road precisely.

6 Conclusions

This paper presented an H_∞ observer to estimate the road profile, using a nonlinear dynamic model of the ER damper. First, the quarter-car system is formulated in a descriptor system in which the nonlinearity coming from the damper model is bounded through a Lipschitz condition. The use of two accelerometers, an H_∞ observer, is designed, providing good estimation results of the road profile (not only ISO but also bump road profiles). The estimation error is minimized by using an H_∞ criteria. Simulation results assess the ability proposed method to estimate the road profile of the semi-active damper.

References

1. ASTM: standard test method for measuring the longitudinal profile of traveled surfaces with an accelerometer established inertial profiling reference (1997)
2. Boyd S, El Ghaoui L, Feron E, Balakrishnan V (1994) Linear matrix inequalities in system and control theory, vol 15. SIAM
3. Darouach M, Boutayeb M (1995) Design of observers for descriptor systems. IEEE Trans Autom Control 40(7):1323–1327
4. Delshad SS, Johansson A, Darouach M, Gustafsson T (2016) Robust state estimation and unknown inputs reconstruction for a class of nonlinear systems: multiobjective approach. Automatica 64:1–7
5. Doumiati M, Martinez J, Sename O, Dugard L, Lechner D (2017) Road profile estimation using an adaptive youla-kučera parametric observer: comparison to real profilers. Control Eng Pract 61:270–278
6. Fialho I, Balas GJ (2002) Road adaptive active suspension design using linear parameter-varying gain-scheduling. IEEE Trans Control Syst Technol 10(1):43–54
7. Fleps-Dezasse M, Svaricek F, Brembeck J (2019) Design and experimental assessment of an active fault-tolerant LPV vertical dynamics controller. IEEE Trans Control Syst Technol 27(3):1267–1274
8. Ha QP, Trinh H (2004) State and input simultaneous estimation for a class of nonlinear systems. Automatica 40(10):1779–1785
9. Healey A, Nathman E, Smith C (1977) An analytical and experimental study of automobile dynamics with random roadway inputs
10. Koenig D, Mammar S (2002) Design of proportional-integral observer for unknown input descriptor systems. IEEE Trans Autom Control 47(12):2057–2062
11. Pham TP (2020) LPV observer and Fault-tolerant control of vehicle dynamics: application to an automotive semi-active suspension system. Ph.D. thesis, Université Grenoble Alpes
12. Pham TP, Sename O, Dugard L (2019) Design and experimental validation of an hinf observer for vehicle damper force estimation. IFAC-PapersOnLine 52(5):673–678
13. Pham TP, Sename O, Dugard L (2019) Real-time damper force estimation of vehicle electrorheological suspension: a nonlinear parameter varying approach. IFAC-PapersOnLine 52(28):94–99
14. Pham TP, Sename O, Dugard L (2019) Unified hinf observer for a class of nonlinear lipschitz systems: application to a real ER automotive suspension. IEEE Control Sys Lett 3(4):817–822
15. Rathai KMM, Sename O, Alamir M (2019) GPU-based parameterized NMPC scheme for control of half car vehicle with semi-active suspension system. IEEE Control Syst Lett 3(3):631–636

16. Savaresi SM, Poussot-Vassal C, Spelta C, Sename O, Dugard L (2010) Semi-active suspension control design for vehicles. Elsevier
17. Solhmirzaei A, Azadi S, Kazemi R (2012) Road profile estimation using wavelet neural network and 7-DOF vehicle dynamic systems. J Mech Sci Technol 26(10):3029–3036
18. Stavens D, Thrun S (2012) A self-supervised terrain roughness estimator for off-road autonomous driving. arXiv:1206.6872
19. Tudón-Martínez JC, Fergani S, Sename O, Martinez JJ, Morales-Menendez R, Dugard L (2015) Adaptive road profile estimation in semiactive car suspensions. IEEE Trans Control Syst Technol 23(6):2293–2305
20. Tudon-Martinez JC, Fergani S, Sename O, Morales-Menendez R, Dugard L (2014) Online road profile estimation in automotive vehicles. In: 2014 European control conference (ECC). IEEE, pp 2370–2375

Control Method for a Wide Object Manipulation for a Team of Collaborative Robots Traveling in Rails

Jirapa Tirasaran and Surat Kwanmuang

Abstract Multirobot systems are used in operations where a single robot is not capable. In this paper, a team of multirobot system is used to transport an object which the carried object is much larger than a single robot can carry. To transport the oversized object smoothly, control and communication between robots is crucial for smooth object manipulation. In this application, the robots move linearly along rails. The different position between robots was measured by using an incremental encoder attached at a support point on the robot to measure the angle between robots. A control feedback method was implement using P control to eliminate the relative position between the master and slave robot. Through experiments, it has been shown that the slave robot can follow the master robot with relative positioning error less than 4 cm, or less than 5° of misalignment.

Keywords Multirobot system · Collaborative robots

1 Introduction

Nowadays, by using robots, humans can reduce their work which has especially long, tedious, heavy, and quite hazardous process. One kind of those works is object transportation, which is tedious work and sometimes, it is also dangerous if carried object is huge, heavy, or unsafe. Mostly, the size of robot for transportation depends on the size of the carried object. However, the limit of the size of robot must be concerned. If huge object is required to carry, multirobot system can be used to solve this problem.

One challenge for manipulating an object using multirobot system is that the team of robots must work collectively together. Any mismatch between robot positions

J. Tirasaran · S. Kwanmuang (✉)
Department of Mechanical Engineering, Chulalongkorn University, Bangkok, Thailand
e-mail: surat.k@chula.ac.th

J. Tirasaran
e-mail: jirapa.trsr@gmail.com

G. Conte and O. Sename (eds.), *Proceedings of 10th International Conference on Mechatronics and Control Engineering*, Lecture Notes in Mechanical Engineering,
https://doi.org/10.1007/978-981-19-1540-6_13

would result in the carried object falling out of the robot or cause damage to the object. In an application where the robots are constrained to move along predetermined axis, this misalignment will cause binding between robots and rails, thus impedes smooth operation. In this paper, we are interested in an application where a team of robot moves along rails on one axis. This application is useful for heavy object transfer between fixed points such as between manufactory stations or between aisles of a warehouse.

Specifically, the contributions of this paper include:

- A purposed method to control a group of robots to move synchronously together using feedback from an optical encoder for relative position measurements.
- An extension, in which, the control gain is a function of robot's velocity to increase sensitivity while still maintaining stability at low speed. Also, the baseline-dependent gain is also explored.
- An evaluation from experiments using real robots operating on various parameters, such as speed and baseline.

2 Related Works

2.1 *Multirobot System*

Multirobot system is used for solving many problems which cannot solve by single robot such as reducing time to explore the environment [1]. In transportation section, there are many cases using multirobot [2–6]. For example, the Mammoet company combined their robots together to carry a huge object and controlled all functions by a single operator, working from one panel [7]. However, if you cannot combine robots to carry an object for your work, you have to make more complicated robots to operate together such as Alonso-Mora and team [8]. To use multirobot system, there are several aspects to concern [9] such as communication [10, 11], object transportation and manipulation, architectures, localization, motion coordination [12, 13], reconfigurable robot, and so on.

There are many multirobot projects made to solve several cases for transporting an object. However, most of all designs are complicated. Whether the sensor, cameras, or many implements are used for controlling movement. So, this paper would like to simplify controlling movement and can apply easier for other projects. One of the interesting related work is Toward the Development of a Material Transport System using Swarms of Ant-like Robots [14]. In this work, multirobot system can transport an object without communication between each other. They used only one leader robot to drag an object to goal and attached force sensor at each robot to sense the movement of an object and follow the leader. Nevertheless, force sensor is quite sensitive. It may be hard for controlling it. Therefore, in this paper, this method will be applied for controlling multirobot systems without communication but changing to use a more suitable and accuracate sensor.

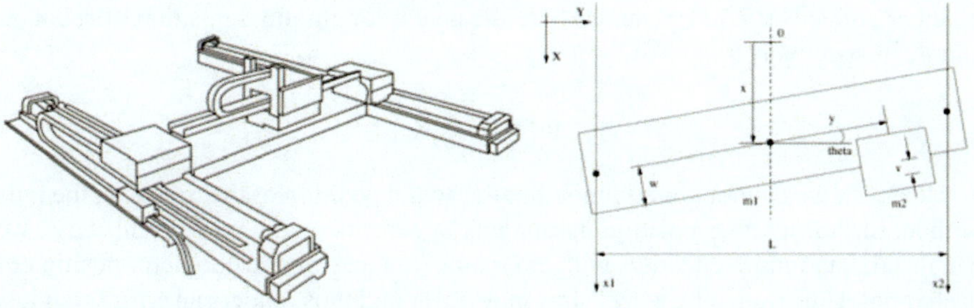

Fig. 1 A gantry system [15] consists of two independent linear actuators. Each actuator has to be controlled in order to move simultaneously

2.2 *Movement*

In our application, each robot has linear motion. The movement of the team of robots to transport an object similar to a gantry which is one type of cartesian robot. In gantry, two motors which are mounted on two parallel slides move a gantry simultaneously [15, 16]. In general, each axis of cartesian system has only one motor to control movement but, in gantry system, there are two motors to control one axis for more an accuracate movement. Principle of gantry robot is each motor sending their position to control movement to get more accuracy. Yaw control is one of the principles of gantry control. It is used to control how bending beam is accepted. That can be converted to how distance the robot follows each other. For robot movement in this study, teams of robots have linear motion and move in tandem like parallel slides to carry an object so yaw control will be applied to use in the robots (Fig. 1).

To apply yaw control, a feedback value was necessary for comparing errors between each robot. To transport smoothly, all robots have to move in unison so the error should be equal to zero.

$$error = x_1 - x_2 \tag{1}$$

when x_1 and x_2 are feedback values from each robot.

3 Proposed Methods

3.1 *Multirobot Position Synchronization Using Wheel Encoders*

For two mobile robots to transport an object along an axis, the relative position along the travel axis is needed to be zero. This indicates that both robots are moving together collectively. Let p_1 and p_2 are the actual longitudinal position of robot #1 and #2, respectively.

$$p_1 - p_2 = 0 \tag{2}$$

However, we cannot directly measure the actual position of the robots on the rails without installing other positioning markers or sensors on the rails. In this case, we can install quadrature encoders with the motors to measure the odometry position of each robot. Unfortunately, wheel slips may occur and thus renders an error ($\epsilon_1$andϵ_2) in wheel position measurement ($\overline{p_1}$ and $\overline{p_2}$)

$$\overline{p_1} = p_1 + \epsilon_1 \tag{3}$$

$$\overline{p_2} = p_2 + \epsilon_2 \tag{4}$$

Thus, by trying to minimize the difference in odometry position,

$$\overline{p_1} - \overline{p_2} = p_1 - p_2 + \epsilon_1 - \epsilon_2 = 0 \tag{5}$$

$$p_1 - p_2 = \epsilon_2 - \epsilon_1 \neq 0 \tag{6}$$

This results in difference in the relative position of the robot due to odometry error. For this reason, we cannot use odometry position to control relative position of two robots without any other sensing method.

3.2 Position Synchronization Using Relative Angle Measurement.

As mentioned above, we need an additional sensing method to measure the relative distance between robots. We installed a quadrature encoder on the object support structure to measure the angle of the object relative to the lateral axis of the robot. This support structure was designed to rotate with the object while allowing the object to slide thus preventing over-constraining the support kinematic (Fig. 2).

The measured angle is a function of robot's relative position and a baseline distance between robots.

$$\theta = \tan^{-1}\left(\frac{p_1 - p_2}{L}\right) \tag{7}$$

Thus, minimizing θ will result in minimizing $p_1 - p_2$ (with scaling factor $1/L$). So, we can then use this angle for a feedback control to minimize relative position.

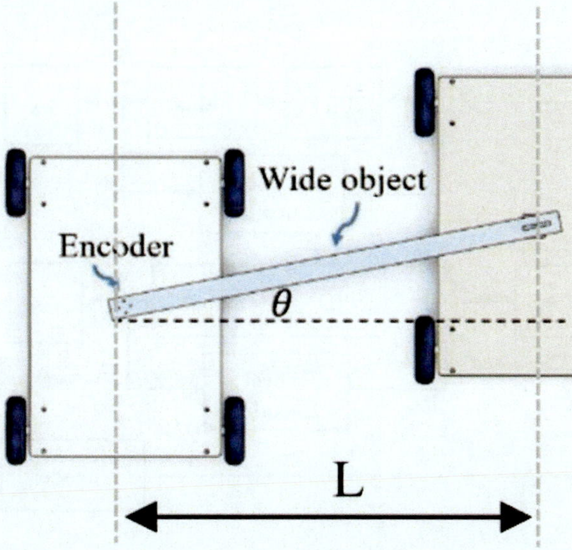

Fig. 2 The angle of the object relative to lateral axis of the robot

The control scheme is the following: The master robot travels with a command speed from telemetry. The slave robot would look for the deviation in the angle measurement and speed up or down trying to minimize this angle. We implemented a P control method to regulate this angle in addition to PID control in velocity mode for both motors (Fig. 3).

$$v_{cmd} = v_{target} + K \cdot (\theta - 0) \tag{8}$$

where

v_{target} is velocity required to transport.
V_{cmd} is a command velocity sent to velocity loop.
K is P control gain.
Θ is converted from the value measured by the encoder.

3.3 Relative Angle Measurement with Velocity-Dependent Gain

The above control method tries to minimize the angle to zero without any other knowledge. This leads to the slowdown of response due to the fact that the angle has to be occurred before the robot can speed up or down. Since both robots have to maintain the same speed to transport the object, we can modify the control law

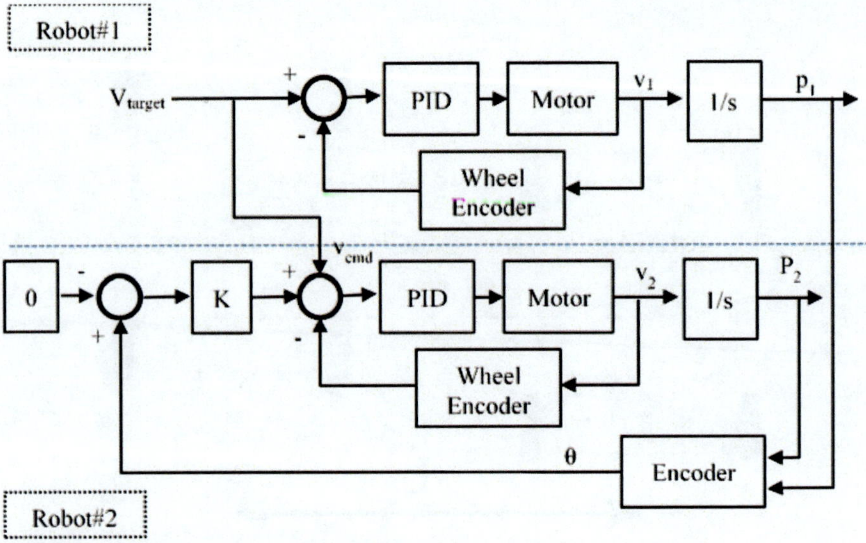

Fig. 3 Diagram of position synchronization using relative angle measurement

specifically the P control gain to be a function of the robot's speed. This increases control sensitivity in a higher speed while maintaining a stable response in lower speed (Fig. 4).

$$v_{cmd} = v_{target} + K \cdot |v| \cdot (\theta - 0) \tag{9}$$

where v is velocity of the slave robot and K is P control gain.

3.4 Relative Angle Measurement with Velocity and Baseline-Dependent Gain

As shown in Fig. 5, the measured relative angle is the same for multiple baseline distances between the robots Z1, Z2 while the relative distance Y1, Y2 is different. The further the two robots are apart (Z), the further the slave robot must travel (Y) to catch up. Thus, when the baseline between two robots is increased, the control gain should be increased in the same fashion.

$$v_{cmd} = v_{target} + k \cdot K_{50} \cdot |v| \cdot (\theta - 0) \tag{10}$$

where K_{50} is the control gain for a baseline at 50 cm. k is the scaling factor between the actual baseline and 50 cm reference baseline.

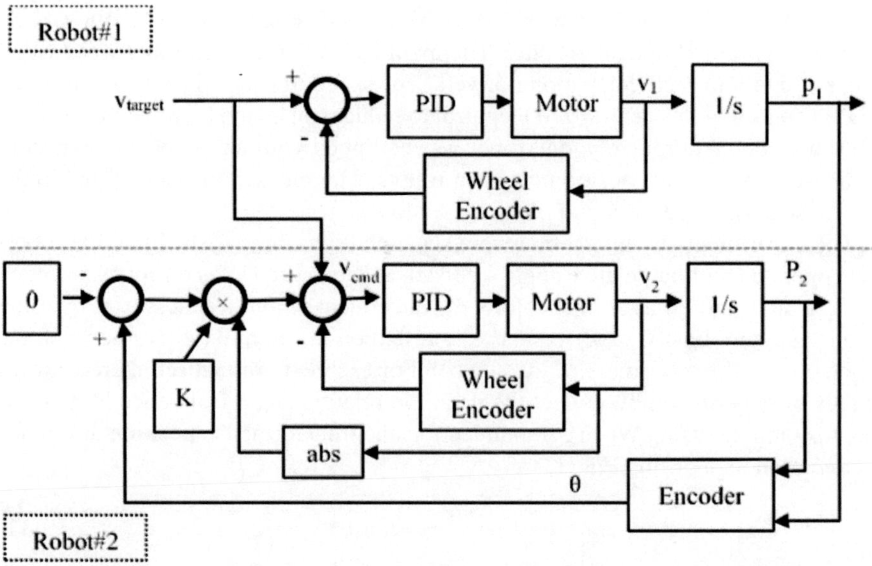

Fig. 4 Diagram of position synchronization using relative angle measurement with velocity-dependent gain

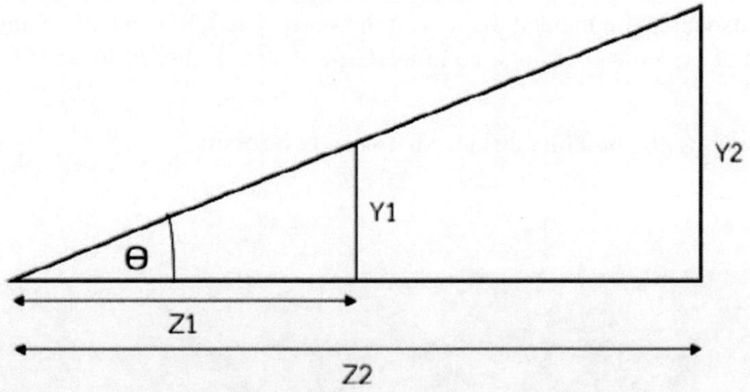

Fig. 5 The relationship between the measured angle, the distance between the robots (Z) and the distance the slave robot needs to travel to be level with the other robot (Y)

4 Experiments and Results

4.1 Experiments

(1) Test of relative position of a multirobot system using encoder feedback

Two-wheeled robots as shown in Fig. 6 were used. Each robot has four wheels with DC motor and optical encoder. The main controller of each robot was ESP32 connected to four VNH5019 motor drivers. The speed of each wheel was controlled by PID control with a centralized P control to reduce mismatch between the position of each wheel. Additionally, one robot was equipped with an incremental encoder to measure the relative position between robots. The measurements then fed to the control scheme as described in previous sections.

The experiments in this section were separated into three parts. First, the robots were operated without relative angle feedback and thus used as a control experiment. Next, relative angle was used to control the movement of multirobot system by using a constant gain. Lastly, experiments were tested on different baselines between the robots to test the robustness of the control. For each experiment set, there were 10 rounds of experiment. To collect the data, the relative angle from encoder (θ) was sent to computer using Wi-Fi communication and transformed to position difference between robots by using Eq. (11)

$$position\_difference = \tan(\theta) * baseline \tag{11}$$

In this test, the baseline between robots was set to 50 cm and a 1-m-wide object represented carried object. Gripper at encoder on robot held the end of object tightly and another side of the object was placed on another robot while allowed to slide freely. At the beginning of each experiment, the relative position was set to zero. Both robots were commanded to move at the same speed. Since the maximum speed of the robot is about 0.48 m/s, command speeds were chosen to be 0.1, 0.2, and 0.3 m/s.

(2) Changing the baseline distance between two robots.

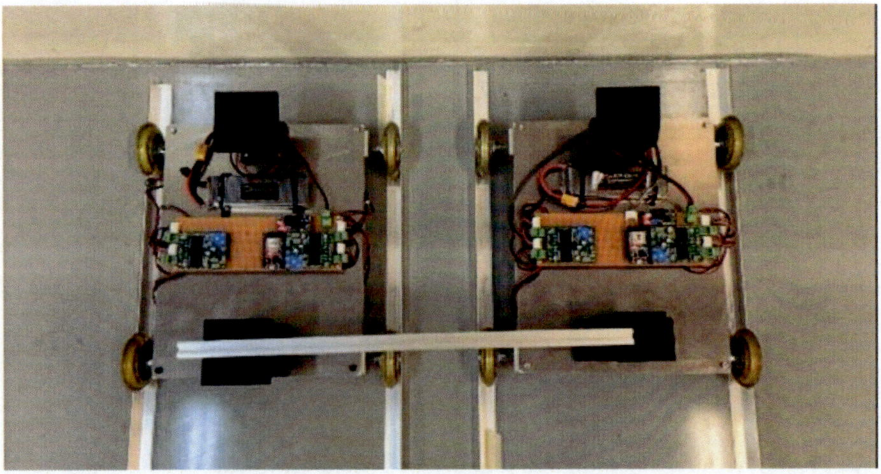

Fig. 6 Two-wheeled mobile robots that were used in the experiments

To verify that the gain selected from the previous tests could apply with other conditions, the carried object was changed to two-meter-wide object and the baseline distance between two robots was increased to 100 and 150 cm. To make the slave robot catch up, the other robot with nearby distance in the previous section, P control gain was also changed to 2- and 3-times as large as the gain at 50 cm baseline, respectively. In this section, there were 10 experiments.

4.2 Results

(1) Test of relative position of a multirobot system using encoder feedback

In the first experiment, there was no feedback in controlling the movement between two robots. The maximum position difference was about 80 cm as shown in Table 1. After that, a feedback control with a constant gain was applied to the movement. This improved the performance of the team of robots. The position difference was reduced significantly to less than 4 cm and the average relative distance was 0.75 cm. as shown in Table 2. However, the result from the test showed that the faster the robots moved, the more likely the position difference was increased. Thus, the gain control was changed to be a function of the robot's velocity. As shown in Table 3, this further improved the performance. The average distance was decreased to 0.35 cm. The control scheme and gain obtained from this experiment was in the next section test since it performed best throughout the experiments (Fig. 7).

(2) Changing the baseline distance between two robots.

Table 1 Position difference between two robots without feedback control

	0.1 m/s		0.2 m/s		0.3 m/s	
	Position difference	±95% confident interval	Position difference	±95% confident interval	Position difference	±95% confident interval
max (cm.)	34.91	±7.47	76.91	±17.55	80.31	±18.09
Average (cm.)	9.29	±5.53	18.79	±11.87	16.56	±13.34

Table 2 Position difference between two robots with constant gain control

	0.1 m/s		0.2 m/s		0.3 m/s	
	Position difference	±95% confident interval	Position difference	±95% confident interval	Position difference	±95% confident interval
max (cm.)	1.18	±0.26	2.55	±0.58	3.78	±0.49
Average (cm.)	0.17	±0.06	0.36	±0.15	0.75	±0.43

Table 3 Position difference between two robots with velocity-dependent gain control

	0.1 m/s		0.2 m/s		0.3 m/s	
	Position difference	±95% confident interval	Position difference	±95% confident interval	Position difference	±95% confident interval
max (cm.)	1.49	±0.29	2.59	±0.57	3.58	±0.85
Average (cm.)	0.11	±0.04	0.22	±0.16	0.35	±0.30

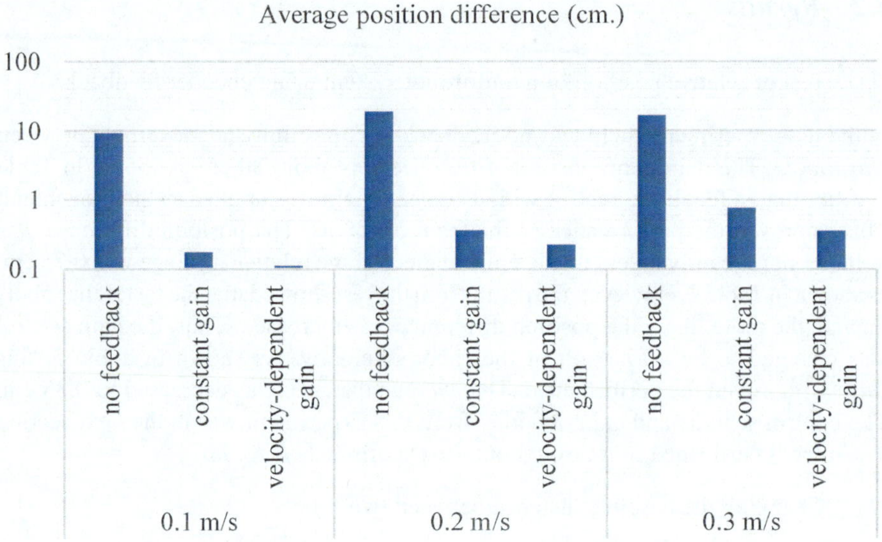

Fig. 7 The comparison of experimental position difference between no feedback control, controlled with a constant gain, and controlled with velocity-dependent gain

Table 4 Position difference between two robots when baseline distance at speed 0.1 m/s

Relative distance between two robots	50 cm		100 cm		150 cm	
	Position difference	±95% confident interval	Position difference	±95% confident interval	Position difference	±95% confident interval
max (cm.)	1.49	±0.29	1.10	±0.22	1.77	±0.35
Average (cm.)	0.11	±0.04	0.13	±0.07	0.12	±0.03

Table 5 Position difference between two robots when baseline distance at speed 0.2 m/s

Relative distance between two robots	50 cm		100 cm		150 cm	
	Position difference	±95% confident interval	Position difference	±95% confident interval	Position difference	±95% confident interval
max (cm.)	2.59	±0.57	2.28	±0.35	2.24	±0.41
Average (cm.)	0.22	±0.16	0.25	±0.20	0.14	±0.07

Table 6 Position difference between two robots when baseline distance at speed 0.3 m/s

Relative distance between two robots	50 cm		100 cm		150 cm	
	Position difference	±95% confident interval	Position difference	±95% confident interval	Position difference	±95% confident interval
max (cm.)	3.58	±0.85	3.06	±0.39	3.30	±0.88
Average (cm.)	0.35	±0.30	0.24	±0.10	0.35	±0.30

As shown in Tables 4, 5, 6, changing baseline distance between two robots with baseline-dependent gain can maintain relatively the same performance. The robots could follow each other with the same performance as in previous tests.

5 Discussion

The result has shown that the multirobot system controlled with velocity-dependent can improve the performance of the movement to transport an object. The difference positions between the two robots was less than 4 cm at all times. Assuming a baseline distance of 1 m, the object would rotate approximately 4.57°, which is many folds improvement from an uncontrolled system. Multirobot can work together to transport large objects smoothly. Moreover, when extending to baseline-dependent gain, the robot can also work at the relatively the same performance with a relative distance less than 4 cm while operating on changing baseline distance.

References

1. Burgard W, Moors M, Stachniss C, Schneider FE (2005) Coordinated multi-robot exploration. IEEE Transactions on robotics, 21(3):376–386
2. Tuci E, Alkilabi MH, Akanyeti O (2018) Cooperative object transport in multi-robot systems: A review of the state-ofthe- art. Frontiers in Robotics and AI, 5: 59
3. Miyata N, Ota J, Arai T, Asama H (2002) Cooperative transport by multiple mobile robots in unknown static environments associated with real-time task assignment. IEEE transactions on robotics and automation, 18(5):769–780
4. Rus D, Donald B, Jennings J (1995 August) Moving furniture with teams of autonomous robots. In Proceedings 1995 IEEE/RSJ International Conference on Intelligent Robots and Systems. Human Robot Interaction and Cooperative Robots (Vol 1, pp 235–242). IEEE
5. Stone P, Asada M, Balch T, D'Andrea R, Fujita M, Hengst B, Wyeth G (2001) Robocup-2000: The fourth robotic soccer world championships. AI magazine, 22(1):11–11
6. Kosuge K, Oosumi T (1996 November) Decentralized control of multiple robots handling an object. In Proceedings of IEEE/RSJ International Conference on Intelligent Robots and Systems. IROS'96 (Vol 1, pp 318–323). IEEE
7. Mammoet (2014). All you need to know about the mammoet self-propelled modular transporter (SPMT). [cited April 29, 2021]. https://www.youtube.com/watch?v=wL3YvOe0ZgE

8. Alonso-Mora J, Baker S, Rus D (2017) Multi-robot formation control and object transport in dynamic environments via constrained optimization. The International Journal of Robotics Research, 36(9):1000–1021
9. Arai T, Pagello E, Parker LE (2002) Advances in multi-robot systems. IEEE Transactions on robotics and automation, 18(5):655–661
10. Schmitt T, Hanek R, Beetz M, Buck S, Radig B (2002) Cooperative probabilistic state estimation for vision-based autonomous mobile robots. IEEE Transactions on robotics and automation, 18(5):670–684
11. Premvuti S, Yuta S (1990 July) Consideration on the cooperation of multiple autonomous mobile robots. In EEE International Workshop on Intelligent Robots and Systems, Towards a New Frontier of Applications (pp 59–63) IEEE
12. Shoval S, Borenstein J (2001 March) Measuring the relative position and orientation between two mobile robots with binaural sonar. In ANS 9th international topical meeting on robotics and remote systems, Seattle, Washington
13. Borenstein J (1994 September) Internal correction of dead-reckoning errors with the smart encoder trailer. In Proceedings of IEEE/RSJ International Conference on Intelligent Robots and Systems (IROS'94) (Vol 1, pp 127–134). IEEE
14. Stilwell DJ, Bay JS (1993 May) Toward the development of a material transport system using swarms of ant-like robots. In [1993] Proceedings IEEE International Conference on Robotics and Automation (pp 766–771). IEEE
15. Teo CS, Tan KK, Lim SY, Huang S, Tay EB (2007) Dynamic modeling and adaptive control of a H-type gantry stage. Mechatronics, 17(7)361–367
16. Li C, Chen Z, Yao B (2018) Advanced synchronization control of a dual-linear-motor-driven gantry with rotational dynamics. IEEE Transactions on Industrial Electronics, 65(9):7526–7535

Design of an Adaptive Fuzzy PID Controller and IoT Platform Implementation for a Teaching Module: Kaplan Turbine

Hernando González Acevedo, Miguel Cordero, Andrés Castro, and Azarquiel Diaz

Abstract The paper presents the dynamic model of a Kaplan turbine coupled to a DC generator, which is part of the H112D teaching module. An adaptive Fuzzy PID controller was designed and it was implemented in a distributed control system, Delta V. For the register of variables was implemented an IoT platform and Heroku was selected as the web server, which supports several programming languages that are used for web application. The control strategy is evaluated by a variable reference signal, leading to limits that saturate the actuator. The nonlinear control technique guarantees a similar transient response at different operating points.

Keywords Kaplan turbine · Adaptive fuzzy PID · IoT platform · Distributed control system

1 Introduction

In order to supply the growing demand for energy in the world, there are various technologies for that generation, from renewable and non-renewable sources, highlighting wind, thermal, tidal, geothermal, hydraulic, and solar energy. Colombia currently has a power generation from seven main types: hydraulic, gas thermal, coal thermal, liquids, gas-liquids, wind, and biomass. According to the Mining-Energy Planning Unit UPME (2015) report, the electricity generated from hydraulic

H. G. Acevedo (✉) · M. Cordero · A. Castro · A. Diaz
Universidad Autónoma de Bucaramanga, Bucaramanga, Colombia
e-mail: hgonzalez7@unab.edu.co

M. Cordero
e-mail: mcordero840@unab.edu.co

A. Castro
e-mail: acastro633@unab.edu.co

A. Diaz
e-mail: adiaz106@unab.edu.co

G. Conte and O. Sename (eds.), *Proceedings of 10th International Conference on Mechatronics and Control Engineering*, Lecture Notes in Mechanical Engineering,
https://doi.org/10.1007/978-981-19-1540-6_14

technology ranks first, with a 70.39% share of total production in the country and followed by gas thermal production with 10.86%.

In Colombia, electricity generation is mainly integrated into two systems, the first is a National Interconnected System, which corresponds mainly to the Andean and Caribbean area, while the second is composed of non-interconnected zones, usually in isolated areas of the country. Most of the electricity generation in the National Interconnected System is derived from hydroelectricity (80%), the remaining 20% is from thermal power plants with a small share of the Small Hydroelectric Power Plants, cogeneration in the agribusiness and a wind farm.

Depending on the conditions where the hydroelectric plant is located, different types of turbines are used. If the water drop is important Pelton turbines are used, if the falls are medium Francis turbines are used, and if the falls are small then Kaplan turbines is the best choice. The dynamics of these turbines have been presented in different papers, using nonlinear models [1, 2], CAE simulation software [3], or using optimization algorithms [4, 5]. The objective of the control system is to maintain a constant speed using techniques such as proportional–integral–derivative control (PID) [6] or robust control [7–9].

2 Dynamic Model

The Autonomous University of Bucaramanga has in its facilities the H112D didactic system, consisting of two reaction turbines: a Francis and Kaplan (Fig. 1), which can

Fig. 1 H112D teaching module

be coupled separately with a DC generator. The flow rate can be adjusted by means of a distributor.

The Kaplan turbine is stored in low head sites with short pipe, therefore, the flow can be considered incompressible. Given this condition, Eq. (1) is set, where h is the turbine head, h_o the initial value of the turbine head, q is the turbine discharge, q_o is the initial value of the discharge, T_w is the time constant, and s is the operator of Laplace. The mechanical torque m is calculated from Eq. (2), where n_t is the turbine efficiency and ω the speed of the shaft that couples the turbine with the generator [10, 11].

$$\frac{q - q_o}{h - h_o} = -\frac{1}{T_w s} \tag{1}$$

$$m = \frac{qhn_t}{\omega} \tag{2}$$

The turbine discharge and efficiency are depending on the head (h), the angular velocity (ω), and the angles of the blades located in the impeller (y_R) and in the distributor (y_W). These two variables are obtained from technical characteristics of the turbine, assuming that the variations in speed are negligible, especially when the unit is connected to the electrical system. The pump head is defined by (3), where h_{char} is the nominal head and $q(y_W, y_R) is$ a nonlinear function that depends on the position of the blades.

$$h = \left(\frac{q}{q(y_w, y_R)}\right)^2 h_{char} \tag{3}$$

Figure 2 presents the blocks diagram of the dynamic model of the Kaplan-type hydraulic turbine coming out from Eqs. (1) to (3), assuming that efficiency, $n_t(y_W, y_R)$, and discharge $q(y_W, y_R)$ are obtained from experimental measurements.

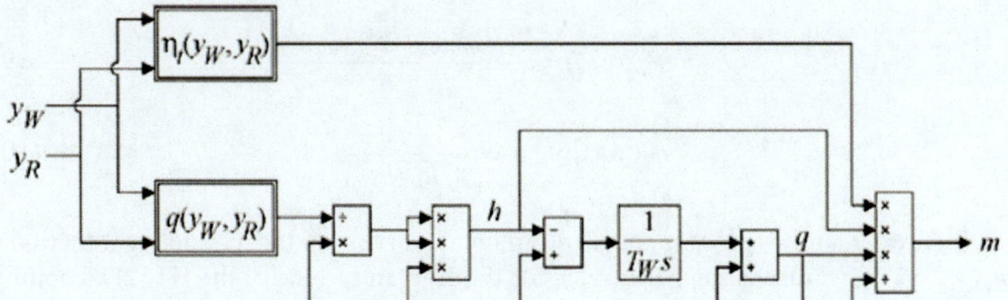

Fig. 2 Block scheme of Kaplan hydraulic turbine model

The dynamic equations of the DC generator are defined by the expressions (4) to (5), where I_a is the armature current, m is the mechanical torque, ω is the mechanical speed, B_m is the viscous coefficient of friction, J_m is the generator inertia, K_I is the torque constant, K_V the electromotive force constant, R_a the armature resistance, L_a the armature inductance, and R the load resistance.

$$m - K_I I_a - B_m \omega = J_m \dot{\omega} \tag{4}$$

$$K_V \omega = L_a \dot{I_a} + (R_a + R) I_a \tag{5}$$

2.1 Linear Model

The system states are $x_1 = \omega$, $x_2 = I_a$ and $x_3 = q - q_o$. The rates of change for each state are defined by the following relationships:

$$\dot{x_1} = \frac{1}{J_m}\left(\left(\frac{(x_3 + q_o)n_t(y_w, y_R)}{x_1}\right)\left(\frac{x_3 + q_o}{q(y_w, y_R)}\right)^2 h_{char} - K_I x_2 - B_m x_1\right) \tag{6}$$

$$\dot{x_2} = \frac{1}{L_a}(K_V x_1 - (R_a + R)x_2) \tag{7}$$

$$\dot{x_3} = -\frac{1}{T_W}\left(\left(\frac{x_3 + q_o}{q(y_w, y_R)}\right)^2 h_{char} - h_o\right) \tag{8}$$

Equilibrium points of states are set for an angular velocity $x_{10} = \omega_o$, therefore, $x_{20} = K_v x_{10}/(R_a + R)$ and $x_{30} = 0$. The equilibrium point of efficiency and discharge are calculated by the following equations:

$$n_{to}(y_{Wo}, y_{Ro}) = \frac{(B_m x_{10} + K_I x_{20})\omega_o}{q_o h_o} \tag{9}$$

$$q_0(y_{Wo}, y_{Ro}) = \frac{q_o}{\sqrt{h_o/h_{char}}} \tag{10}$$

Figures 3 and 4 present the experimental curves of the efficiency functions, $n_t(y_w, y_R)$, and discharge, $q(y_w, y_R)$, for the Kaplan turbine of the H112D teaching system, in which, the position of the blades is normalized. Table 1 shows the value of the parameters of the dynamic model of the turbine coupled to the DC generator.

Assuming $\omega_o = 120\,\frac{\text{rad}}{\text{s}}$, it's obtained $x_{20} = 4A$, $n_{to}(y_{Wo}, y_{Ro}) = 0.5405$ and $q_0(y_{Wo}, y_{Ro}) = 134.72\text{m}^3/h$. The linear model in state space is

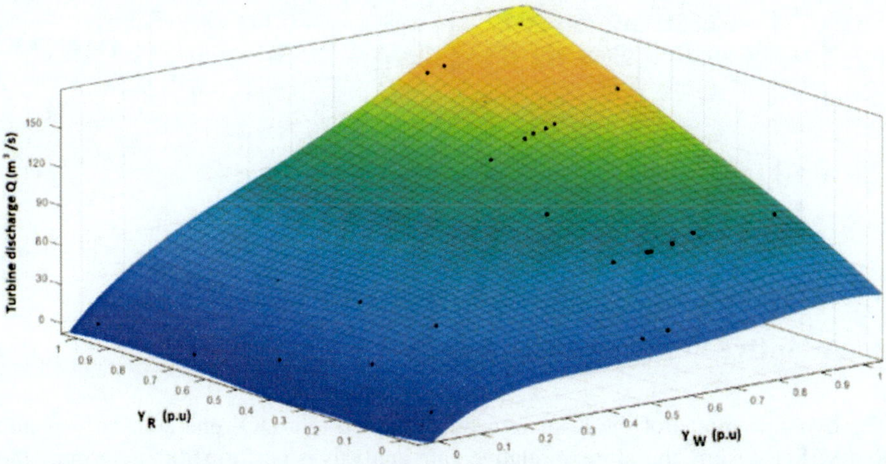

Fig. 3 Measurements data and approximation $q(y_w, y_R)$

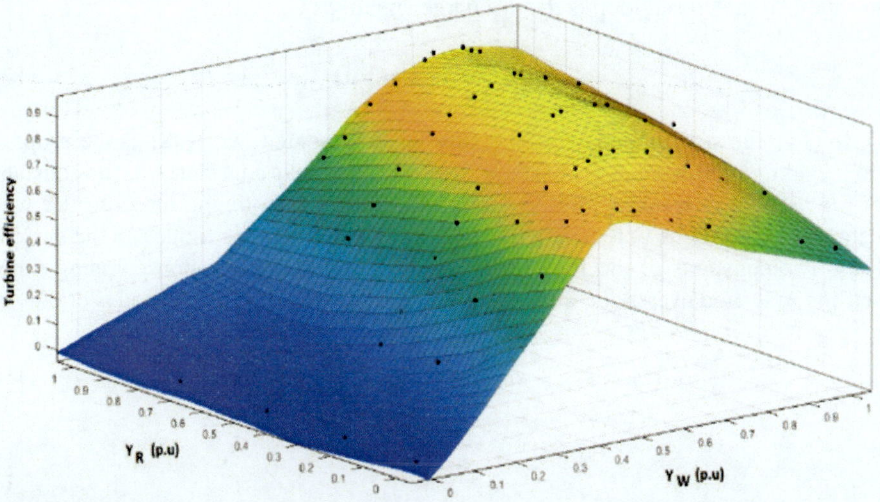

Fig. 4 Measurements data and approximation $n_t(y_W, y_R)$

Table 1 Nominal values and characteristics of teaching module: Kaplan Turbine

Parameter	Value	Parameter	Value
J_m	$0.4\,\mathrm{Kgm}^2$	R	20Ω
B_m	$0.001\mathrm{Nm/rad/s}$	h_o	$8\,\mathrm{m}$
K_I	$0.94\mathrm{Nm/A}$	q_o	$110\,\mathrm{m}^3/\mathrm{h}$
K_V	$0.92\,V/\mathrm{rad}$	h_{char}	$12\,\mathrm{m}$
L_a	$0.0114H$	T_w	$0.2\,\mathrm{s}$
R_a	7Ω		

$$\begin{bmatrix} \dot{x}_1 \\ \dot{x}_2 \\ \dot{x}_3 \end{bmatrix} = \begin{bmatrix} -0.0851 & -2.35 & 0.27 \\ 80.7 & -2368.4 & 0 \\ 0 & 0 & -0.727 \end{bmatrix} \begin{bmatrix} x_1 \\ x_2 \\ x_3 \end{bmatrix} + \begin{bmatrix} -0.147 & 18.33 \\ 0 & 0 \\ 0.594 & 0 \end{bmatrix} \begin{bmatrix} q(y_w, y_R) \\ n_t(y_w, y_R) \end{bmatrix}$$

$$y = \begin{bmatrix} 1 & 0 & 0 \end{bmatrix} \begin{bmatrix} x_1 \\ x_2 \\ x_3 \end{bmatrix} \tag{11}$$

3 Adaptive Fuzzy PID Controller

The dynamic model of the Kaplan turbine coupled to a DC generator corresponds to a MISO system, therefore, a relative gain analysis is performed to determine the appropriate actuator to regulate the mechanical speed, Eq. 12, where K is the DC gain of the transfer function matrix. According to (12), the recommended variable to establish the closed loop is the discharge signal, $q(y_W, y_R)$.

$$RGA = K * \left(K^{-1}\right)' \tag{12}$$

In (13), the structure of the PID controller is presented, where K_P is the proportional gain, K_I is the integral constant, K_D is the derivative constant, N is the constant of the pseudo derivator, and T_s is the sample period. The tuning of controller was determined using the pole location technique, given the linear model on the system [12]. The constants of the PID controller and the unit step response (rising time, settling time, and overshoot) are defined in Table 2. The sampling period is of 0.01 s.

$$G(z) = k_{po} + k_{io} T_s \frac{1}{z-1} + k_{do} \frac{N}{1 + N.T_s \frac{1}{z-1}} \tag{13}$$

3.1 Fuzzy PID Controller

Fuzzy adaptive PID controller applies fuzzy control rules to modify PID parameters online using error (E) and change-in-error (EC) as inputs, which can meet the request of E and EC in different time for self-tuning PID parameters. Figure 5 illustrates the

Table 2 Constants of the PID controller

k_{po}	k_{io}	k_{do}	N	Rise time	Settling time	Overshoot
0.0074	0.00254	−0.0065	3.33	7.19 s	23.1 s	7.71%

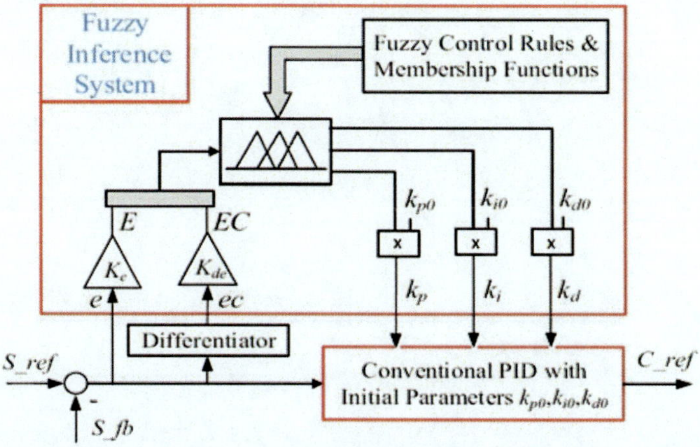

Fig. 5 Block diagram of an adaptive Fuzzy PID controller

block diagram of an adaptive fuzzy PID controller. It contains a conventional PID and a fuzzy inference system. The design work includes the following aspects: (1) the initial parameters k_{po}, k_{io}, and k_{do} of conventional PID; (2) membership functions of fuzzy inference system; (3) fuzzy control rules; and (4) the input scaling factors k_e, k_{de}. The objective of the constants k_e and k_{de} is to normalize the input variables to the fuzzy system in between a range of -1 and 1 [13].

The constants of the PID controller are defined by the next relations: $K_p = k_1 k_{p0}$, $K_i = k_2 k_{i0}$, and $K_d = k_3 k_{d0}$, where k_1, k_2, and k_3 are outputs of the fuzzy controller and remain within a range from 0 to 2. For each of the linguistic variables were defined seven membership functions: Positive big (PB), positive medium (PM), positive small (PS), zero (ZO), negative small (NS), negative medium (NM), and negative big (NB). The membership functions defined for the input and output signals can be observed in the Fig. 6, they are conformed by triangular functions in the center and trapezoid functions in the edges.

3.2 *Design of Fuzzy Control Rules*

The fuzzy control rule is the summary of the system control experience that captures the expert's knowledge about how to control the plant, which directly affects the quality of control system. In this paper, fuzzy control rules are designed with these tune principles as follows:

- The constants of the PID controller remain close to the tuned values when small variations appear in the angles in relation with the reference signal.
- When $E > 0$, $EC < 0$ and E is large, in order to maintain a rapid response and make the absolute of error reduce in the maximum speed, K_p should be bigger

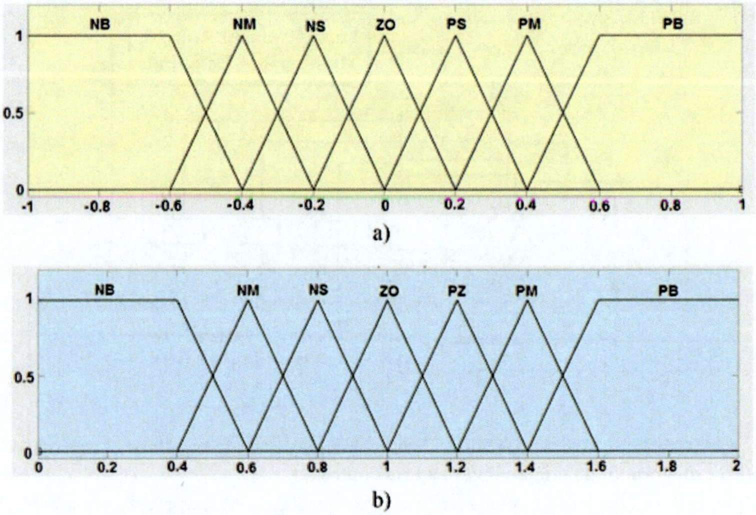

Fig. 6 Membership functions for the fuzzy sets. **a** Linguistic variable: error (E) and error rate of change (EC) **b** Linguistic variable of the output: k_1, k_2, and k_3

and K_i, K_d should be smaller; with the decrease of E, to prevent the overshoot be too big, K_d should be added andK_p, K_i should be diminished, evenly, K_i should be cancelled.

- When $E < 0$, $EC > 0$, the system is over steady state and the error is increasing, in order to decrease the overshoot, K_d should be bigger and K_i, K_p, should be smaller.
- When $E < 0$, $EC > 0$, the system is tending steady state, K_p should be taken a bigger value in order to promote the response speed and then access to steady state quickly; K_d should be added to decrease the overshoot; K_i should be diminished to avoid oscillation caused by integral overshoot.
- When $E > 0$, $EC > 0$, the overshoot of the system is negative. When the error is increasing, K_d should be taken a bigger value, once the error reaches the maximum and the system is tending steady state, K_d should be decreased, and K_p, K_i should be increased.

According to a big number of experimental tests, the rule base for the fuzzy controller can be observed in Table 3 (proportional gain), the Table 4 (integral gain), and the Table 5 (derivative gain).

Table 3 Fuzzy rule base for K_P

		E						
		NB	NM	NS	ZO	PS	PM	PB
EC	NB	PB	PB	PM	PM	PS	ZO	ZO
	NM	PB	PB	PM	PS	PS	ZO	NS
	NS	PM	PM	PM	PS	ZO	NS	NS
	ZO	PM	PM	PS	ZO	NS	NM	NM
	PS	PS	PS	ZO	NS	NS	NM	NM
	PM	PS	ZO	NS	NM	NM	NM	NB
	PB	ZO	ZO	NM	NM	NM	NB	NB

Table 4 Fuzzy rule base for K_I

		E						
		NB	NM	NS	ZO	PS	PM	PB
EC	NB	NB	NB	NM	NM	NS	ZO	ZO
	NM	NB	NB	NM	NS	NS	ZO	ZO
	NS	NB	NM	NS	NS	ZO	PS	PS
	ZO	NM	NM	NS	ZO	PS	PM	PM
	PS	NM	NS	ZO	PS	PS	PM	PB
	PM	ZO	ZO	PS	PS	PM	PB	PB
	PB	ZO	ZO	PS	PM	PM	PB	PB

Table 5 Fuzzy rule base for K_D

		E						
		NB	NM	NS	ZO	PS	PM	PB
EC	NB	PS	NS	NB	NB	NB	NM	PS
	NM	PS	NS	NB	NM	NM	NS	ZO
	NS	ZO	NS	NM	NM	NS	NS	ZO
	ZO	ZO	NS	NS	NS	NS	NS	ZO
	PS	ZO	ZO	ZO	ZO	ZO	ZO	ZO
	PM	PB	PS	PS	PS	PS	PS	PB
	PB	PB	PM	PM	PM	PS	PS	PB

3.3 Experimental Validation

The control strategy was programmed in the software DeltaV using a tool called Control Studio, which presents blocks with specific functions that are structured in a similar way of a flux diagram. This tool has general purpose blocks, these

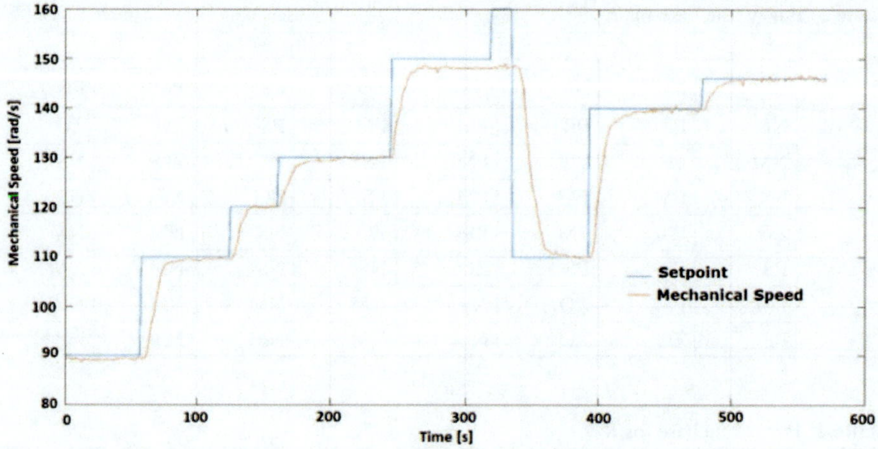

Fig. 7 Transient response of the mechanical speed for a variable setpoint

blocks are programmed in C language, allowing the implementation of fuzzy PID algorithm [14]. The position of the turbine blades (y_w, y_R) is calculated from the experimental data in Figs. 3 and 4, performing a polynomial approximation, known the value of $q(y_w, y_R)$ and $n_t(y_W, y_R)$. Figure 7 shows the response transitory of turbine mechanical speed for a variable setpoint, around the point of operation $\omega_o = 120\ \frac{\text{rad}}{\text{s}}$. The settling time is around of 25 s and overshoot is null. The fuzzy controller allows a fast return when saturating the actuator, $300\text{s} < t < 350\text{s}$, also the speed is not equal to the reference when changes are equal to $\pm 30\text{rad/s}$.

$$y_w = \sum_{k=o}^{r} a_k q(y_w, y_R)^k + \sum_{k=o}^{p} b_k n_t(y_W, y_R)^k$$

$$y_R = \sum_{k=o}^{m} c_k q(y_w, y_R)^k + \sum_{k=o}^{n} d_k n_t(y_W, y_R)^k \quad (14)$$

4 Implementation of IoT Platform

To send the process variables to the cloud, first a serial communication card with the Modbus RTU communication protocol is used. The DeltaV™ S-series Serial Interface provides a connection between the DeltaV system and other devices. This card communicates with an Arduino Uno, the MAX485 module has the function of adopting half-duplex communication converting TTL level into RS-485 level. NodeMCU ESP32 allows WIFI communication, it communicates with Arduino

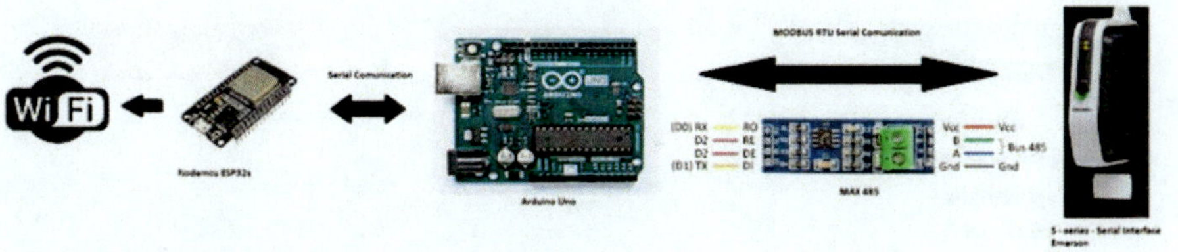

Fig. 8 Schematic of communication diagram

through serial communication. Figure 8 shows the communication diagram that allows making the Modbus RTU communication to Wi-Fi communication.

The system that handles data, transactions, and database problems is known as the Database Management System (DBMS). For this project, MongoDB was selected because it allows users to manage a free storage plan. MongoDB, is a NoSQL database, which stores data in flexible documents with a structure similar to a JSON file, which means that the fields may vary from one document to another and the structure may vary over time [15].

Heroku was selected as web server, it allows to implement, run, and manage web applications, in different programming languages: Ruby, Node.js, Java, Python, Clojure, Scala, Go, and PHP. Heroku manages a free development plan, which allows to use a web application for 550 free hours per month, when the application is not in use for more than 30 min, it is suspended and stops counting hours of use. Figure 9 shows the control panel, the user can manipulate the setpoint, also activate and deactivate the valves located in the process. The option *"Descargar Histórico"* allows download the process variables in an excel file with a sampling time of one minute.

One essential issue in cloud computing is data security, which is handled using cryptography methods. To enhanced data security and privacy, researcher combines AES 256 (Advanced Encryption Standard), IDA (Information Dispersal Algorithm) and SHA 512 (Secure Hash Algorithm). During the process encoding, the original data is encrypted using AES 256 algorithm, encryption generated is randomly by IT manager. Then the encrypted file is divided into several separate files. During the decoding process, verification stages are observed first. After that, reconstruct the encrypted data using IDA, then reconvert original data using AES 256 to get original data. The result shows, average execution time higher when decoding time process is verification 1.453. This topic has been analyzed by several researchers and corresponds to a second stage of the project to develop and implement encryption algorithms for both the MongoDB database and the Heroku web server [16].

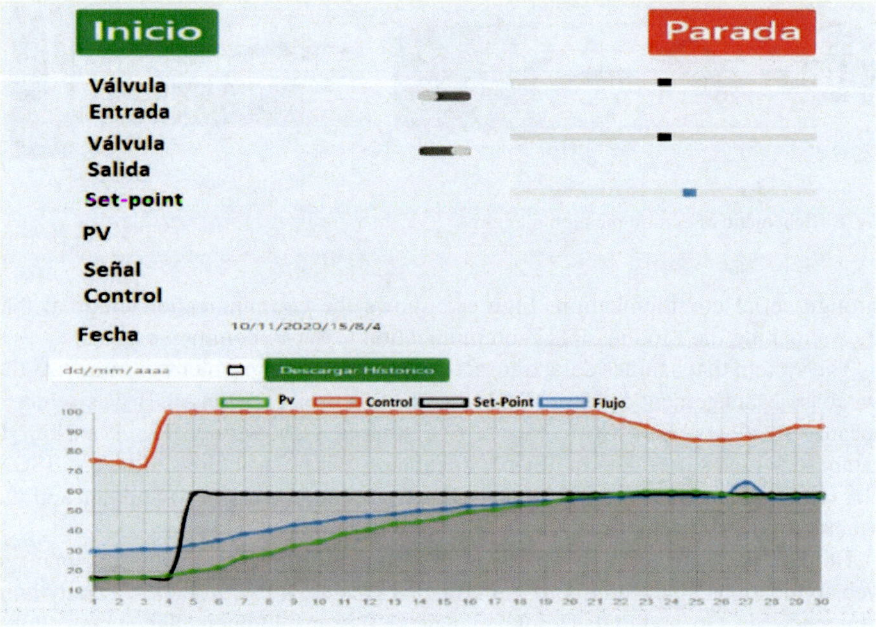

Fig. 9 Human–machine interface

5 Conclusions

This paper presents a mathematical model of a Kaplan-type turbine and the design of an adaptive fuzzy PID. The nonlinear control technical, based on artificial intelligence, guarantees the stability of the system for different operating points, maintaining the same transient characteristics, overshoot and settling time, for each change of the reference signal. Today's existing tools allow implementing IoT platforms of easy and economical way, guaranteeing access to information in real time.

References

1. Izena A, Kihara H, Shimojo T, Hirayama K, Furukawa N, Kageyama T, Goto T, Okamura C (2006) Practical hydraulic turbine model. IEEE Power Eng Soc General Meeting. ISBN: 1–4244–0493–2. https://doi.org/10.1109/PES.2006.1709306
2. Choo YC, Muttaqi KM, Negnevitsky M (2007) Modelling of hydraulic turbine for dynamic studies and performance analysis. Australasian Universities Power Engineering Conference, Dec 2007. ISBN: 978–0–646–49488–3. https://doi.org/10.1109/AUPEC.2007.4548104
3. Stipisei L, Kosi D, Kurtela M (2012) Design and simulation of turbine governing hydraulic system with FluidSIM software. In: Proceedings of the 35th international convention MIPRO, May 2012. ISBN: 978–953–233–068–7

4. Lei G (2017). Application improved particle swarm algorithm in parameter optimization of hydraulic turbine governing systems. In: IEEE 3rd information technology and mechatronics engineering conference (ITOEC), Oct 2017. ISBN: 978–1–5090–5363–6. https://doi.org/10.1109/ITOEC.2017.8122532
5. Zhang Y; Zhou J, Xu Y, Zheng Y (2018) Parameter identification of hydraulic turbine governing system using white-box mapping model and RBF neural network. In: Eighth international conference on instrumentation & measurement, computer, communication and control (IMCCC), July 2018. ISBN: 978–1–5386–8246–3. https://doi.org/10.1109/IMCCC.2018,00030
6. Lei G (2017) Application of improved PID control technology in hydraulic turbine governing systems. In: 3rd international conference on control, automation and robotics (ICCAR), April 2017. ISBN: 978–1–5090–6088–7. DOI: https://doi.org/10.1109/ICCAR.2017.7942702
7. Chen L-Y, On H (2010) Robust control for hydraulic servo system of steam turbine. In: International conference on electrical and control engineering, June 2010. ISBN: 978–1–4244–6881–2. https://doi.org/10.1109/iCECE.2010.704
8. Heng Q, Lu J, Lu Y (2012) Robust control of hydro turbine speed governor. In: Proceedings of the 10th world congress on intelligent control and automation. July 2012. ISBN: 978–1–4673–1398–8. https://doi.org/10.1109/WCICA.2012.6358326
9. Guo C, Li Z, Du C; Fengyang Wang QFT (2012) Synchronous control system of ring gate for hydraulic turbine. In: Asia-Pacific power and energy engineering conference. Mar 2012. ISBN: 978–1–4577–0547–2. https://doi.org/10.1109/APPEEC.2012.6306999.
10. Brezovec M, Kuzle I, Tomisa T (2006) Nonlinear digital simulation model of hydroelectric power unit with Kaplan turbine. IEEE Trans Energy Convers 21(1). ISSN: 0885–8969. https://doi.org/10.1109/TEC.2005.847963
11. Menarin H, Costa H, Minetto G, Pereira R, Finardi E, Weiss L (2019) Dynamic modeling of Kaplan turbines including flow rate and efficiency static characteristics. IEEE Trans Power Syst 34(4). ISSN: 0885–8950. https://doi.org/10.1109/TPWRS.2019.2899815
12. Katsuhiko Ogata (1995) Discrete-time control systems, 2nd edn. Edition Published by Prentice Hall. ISBN 0–13–034281–5
13. González H, Arizmendi C, Garcia J, Angulo A, Herrera C (2018) Design and experimental validation of adaptive fuzzy PID controller for a three degrees of freedom helicopter. In: IEEE international conference on fuzzy systems (FUZZ-IEEE), July 2018. ISBN:978–1–5090–6021–4. https://doi.org/10.1109/FUZZ-IEEE.2018.8491512
14. Emerson Homepage (2021). https://www.emerson.com/en-us/catalog/deltav-ve2161. Aaccessed 01 June 2021
15. Patil M; Hanni A; Tejeshwar CH; Patil P (2017) A qualitative analysis of the performance of MongoDB vs MySQL database based on insertion and retriewal operations using a web/android application to explore load balancing—Sharding in MongoDB and its advantages. In: International conference on I-SMAC (IoT in Social, Mobile, Analytics and Cloud) (I-SMAC), Feb 2017. ISBN:978–1–5090–3244–0. https://doi.org/10.1109/I-SMAC.2017.8058365
16. Kumar J; Garg V, Security analysis of unstructured data in NOSQL MongoDB database. In: International conference on computing and communication technologies for smart nation (IC3TSN). ISBN:978–1–5386–0628–5. https://doi.org/10.1109/IC3TSN.2017.8284495. Lee B; Kusuma E, Farid M (2018) Data security in cloud computing using AES under HEROKU cloud. In: 27th wireless and optical communication conference (WOCC), June 2018. ISBN:978–1–5386–4960–2. https://doi.org/10.1109/WOCC.2018.8372705

Data Analysis and Predictive Control for a Pusher-Type Billets' Reheating Furnace

Silvia Maria Zanoli, Crescenzo Pepe, Giacomo Astolfi, Angelo Di Carlo, and Rodolfo Mathis

Abstract In this paper, the reheating process and the initial milling process, i.e. the roughing phase of steel billets, are analyzed. An Advanced Process Control has been developed and installed on pusher-type billets' reheating furnace of an Italian steel industry. Data analysis is focused on the proof that the installed measurement tools are reliable and suitable for control purposes. A Model Predictive Control approach is exploited based on an adaptive model of the plant that takes into account the validated measurements. The main purpose of the work is to find a model capable of correlating the absorptions of the first six rolling stands with process variables and field measurements taking into account the different operating conditions of the furnace. The designed control system has been installed on the real plant providing a significant support for plant operators and optimizing the furnace conduction. The developed system received Industry 4.0 certification.

Keywords Advanced process control · Pusher type billets' reheating furnace · Model predictive control

1 Introduction

The reheating process represents a key phase in the steel industry production chain: raw materials derived from a continuous casting need to be suitably reheated in order to satisfy the specifications for the desired rolling procedure [1]. In the last years, reheating furnace control and optimization has attracted the attention of process engineers and researchers, due to the challenging research of advantageous trade-offs between energy saving and guaranteeing the required temperature profile of the reheated pieces [2–5].

S. M. Zanoli (✉) · A. Di Carlo · R. Mathis
Dip. Ingegneria dell'Informazione, Univ. Politecnica delle Marche, 60131 Ancona, Italy
e-mail: s.zanoli@univpmi.it

C. Pepe · G. Astolfi
Alperia Bartucci Spa, 37038 Soave, Italy

G. Conte and O. Sename (eds.), *Proceedings of 10th International Conference on Mechatronics and Control Engineering*, Lecture Notes in Mechanical Engineering,
https://doi.org/10.1007/978-981-19-1540-6_15

Hot rolling mills require billets to be reheated to temperatures of around 1200 °C, which is suitable for plastic deformation of steel [6]. In the present paper, an Advanced Process Control (APC) system for controlling and optimizing a pusher-type billets' reheating furnace located in an Italian steel industry is described and its control performances with respect to the rolling phase are analyzed.

The billet reheating and roughing phases are depicted in Fig. 1. The billets enter the furnace from the right side and their input temperature can vary in the range 0–910 °C.

The entry temperature of the billets into the furnace is measured by two optical pyrometers located near the furnace inlet, and it is not known a priori. The furnace is composed by four regions (see Fig. 2 and Table 1); the maximum number of billets that the furnace can contain is 128, and each region is characterized by a different storage capacity (see Table 1). Each region is characterized by one or more sets of burners that regulate the furnace combustion. Regions 1–3 are characterized by one set of burner each, so they are also called zones (zone 1, zone 2, zone 3);

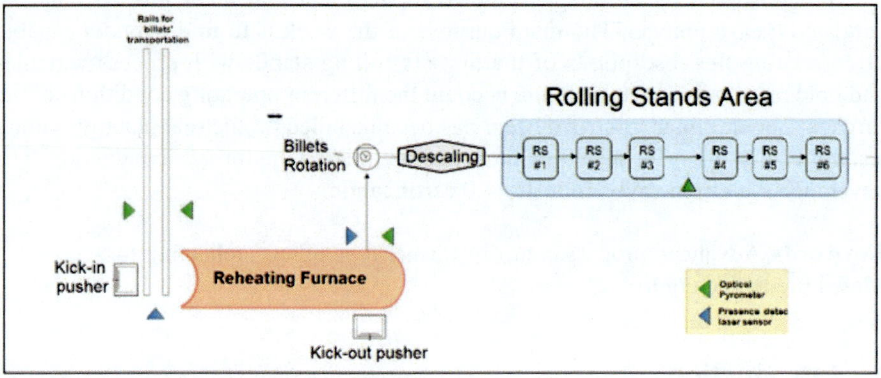

Fig. 1 A scheme of the reheating and roughing phase

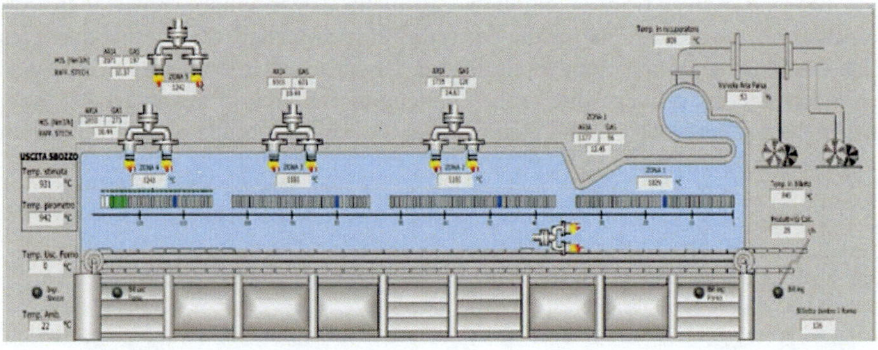

Fig. 2 The pusher type-reheating furnace

Table 1 Furnace zones features

Furnace zone	Capacity (billets)	Length (m)	Temperature setpoint range (°C)
Zone #1	35	5495	825–1020
Zone #1	37	5735	925–1090
Zone #1	31	4710	1025–1180
Zone #1	25	3925	1115–1240

region 4 is characterized by two zones (zones 4–5) that allow reheating billets' head and tail at different temperatures. The temperature of each zone is regulated by a PID (Proportional-Integral-Derivative) controller; typical setpoint ranges have been reported in Table 1. The billets that enter the furnace have a section equal to 150 mm × 150 m, a length equal to 12.2 m and a mass equal to 2155 kg. The furnace movement is managed by pushers that push the billets and translate them along the furnace length; the furnace movement time (and the furnace production rate) can vary based on the final product that will be obtained after the rolling phase. After the furnace exit (Fig. 1, left side), the billets enter the first rolling mill stand: a significant measurement of the effort and of the efficiency of this phase is given by the absorption of each stand. Obviously, the efficiency of the rolling phase is strictly related to the efficiency of the previous phase, i.e. the reheating: in case billets entering the rolling mill stands are too cold, a much higher energy consumption is required to process the billets and, furthermore the rolling mill stands can break leading to significant damages and maintenance costs; on the other hand, overheated billets can lead to inter-locking of the rolling mill stands and/or to an excessive fuel specific consumption. In order to create a temperature model for each billet (needed for control purposes), an optical pyrometer has been placed between the third and the fourth rolling mill stand.

In the current conduction, the predictive control system (MPC) monitors the temperatures of the various areas of the furnace in order to provide billets with a suitable heating temperature profile and to rise the billets' temperature at the furnace outlet to a specific set-point value. This set-point value is currently set by process operators and is determined by their experience and depending on the type of production, so as to keep the energy absorption of the rolling stands within a given range.

This work is aimed at analyzing the absorption of the first set of six rolling stands (roughing phase) downstream of the reheating furnace. In particular, the purpose is to find a model capable of correlating the absorptions of the first six rolling stands with process variables and field measurements taking into account the different operating conditions of the furnace.

The paper is organized as follows: Sect. 2 describes data analysis and process modeling, while Sect. 3 reports the control strategy. Field results are summarized in Sect. 4; conclusions are reported in Sect. 5.

2 Data Analysis and Process Modeling

Process and MPC controller data from log files for the period February–April, 2019 are analyzed. From the knowledge of the heating process and of the rolling mill stands, the following variables have been selected: the inlet temperature, the outlet temperature, the filtered inlet temperature, the absorption of the first six rolling stands, the total permanence time for each billet in the furnace. The steel plant production is very differentiated and in the considered period up to more than 200 different final products were produced; for example, steel rods production can differ in the final diameter (typically from 8 to 40 mm), the steel quality and final external finishing. The present work has been focused on the production of 12 mm steel rods, called GP12. The choice was motivated by the fact that this product was one of the most frequently produced in the considered period and its production was characterized by long and continuous production intervals. Figure 3 shows the hot rolling mill stand absorptions versus the billet outlet temperature for each of the six rolling mills for the considered period.

From a visual inspection, it is clear that a simple linear regression model cannot suitably predict the rolling mill stand absorptions Thus, data partitioning has been

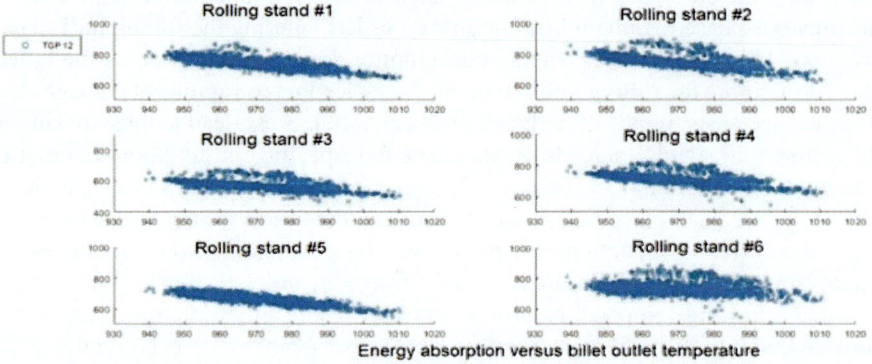

Fig. 3 Billet outlet temperature versus rolling mill stand absorption

Table 2 Data partitioning set

Dataset	Inlet temperature range (°C)
Class 1	801–
Class 2	601–800
Class 3	301–600
Class 4	151–300
Class 5	0–150

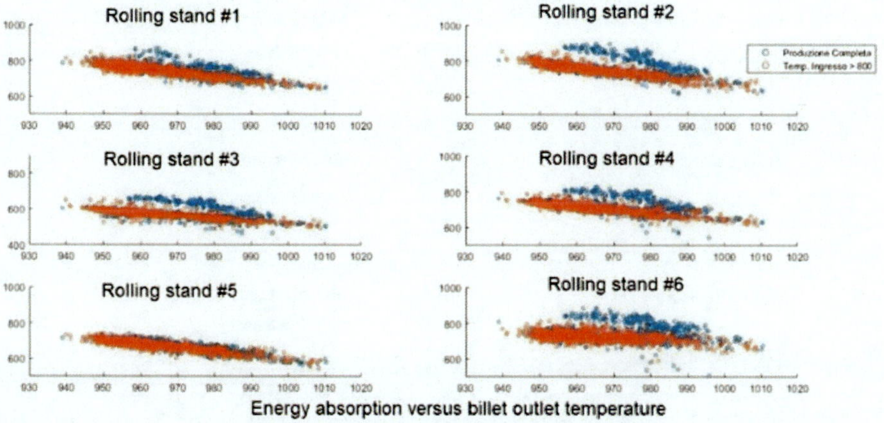

Fig. 4 Rolling mill stand absorption versus billet outlet temperature: red plot refers to inlet temperature higher than 800 °C

performed in order to group billet with similar behavior with respect to energy absorption. As a starting point, five grouping sets have been considered; Table 2 shows the input temperature range that characterizes each dataset.

The analysis of the data carried out initially on all the billets and subsequently by grouping data by temperature subintervals revealed the fact that the colder the billets are at the entrance of the furnace, the greater their variability around absorption expected values. This can be observed from the inspection of plot reported in Figs. 3, 4 and 5. In Fig. 4, data that refer to billets that have entered the furnace with a temperature higher than 800 °C are plotted in red while all the others in blue. The plot shows that considering billets with the same outlet temperature, the ones that entered the furnace at a high temperature require a lower energy to be processed at the rolling mill. Beside there is a marked linear relationship between outlet temperature and absorption.

On the other hand, a more blurred behavior can be observed in the scatterplot in Fig. 5 that refers to billets that entered the furnace at temperature lower than 600 °C. The same behavior occurred even considering smaller data sets (see Table 2) here not reported for brevity. Furthermore, Fig. 5 shows that when considering the filtered inlet temperature similar behavior is observed. This motivated the application of multivariate regression techniques so to predict the energy absorption. From the analysis of the system behavior, four regressors (prediction variables) were chosen to predict the rolling mill absorption of each of the six stands:

$$\hat{y}_i = \beta_{0i} + \beta_{1i} * x_1 + \beta_{2i} * x_2 + \beta_{3i} * x_3 + \beta_{4i} * x_4 \; i = 1..6 \quad (1)$$

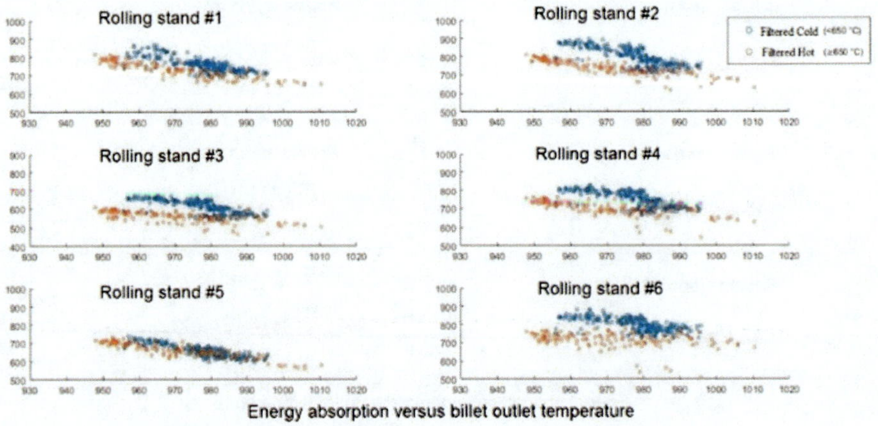

Fig. 5 Rolling mill stand absorption versus billet outlet temperature: the data set refers to billet input temperature lower than 600 °C. Red plots refer to inlet *filtered* temperature higher than 650 °C

where y_i is the energy absorption of the ith rolling stand, $x_1$is the outlet temperature, x_2 is the filtered inlet temperature, x_3 is the total reheating time and x_4 is the inlet temperature.

From a deeper analysis based on F-test and the p-value computation, it was concluded that removing the inlet temperature as regressor, the performances of the reduced regression were not significantly altered. Consequently, only the first three prediction variables were adopted for the estimation of the rolling stands absorption.

In Fig. 6, the absorption prediction computed by the multivariate regression model is compared with real data. The performances are satisfactory; besides, it can be noted that the use of the filtered input billet temperature in the regressor gives results more reliable than the ones obtained with the real input temperature. Given the fact that the considered reheating furnace is a pusher type where billets are attached one to another, this result has a physically explanation: few cold billets within many

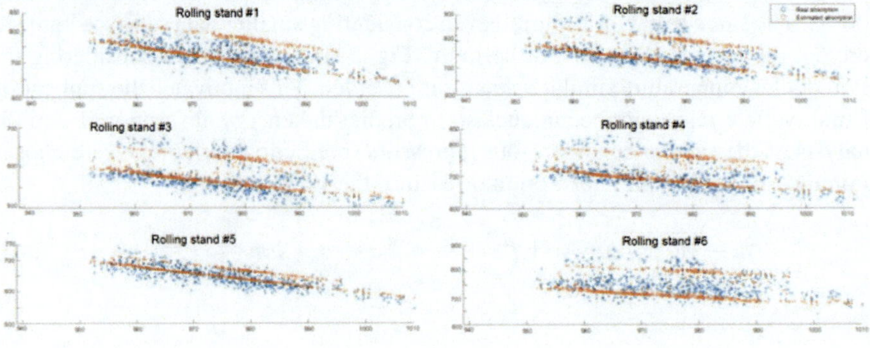

Fig. 6 Multivariate regression model performances

hot billets benefit from the neighboring billets' heat and, for APC purpose, can be considered as hot entering billets.

The previous result has motivated the development of a suitable filtering procedure of the inlet temperature to be used for control purpose that takes into account billet furnace billets' loading sequence.

3 APC System Design

In order to design a model-based controller for the automatic management of the zone temperatures setpoint, the Manipulated Variables (MVs), the Disturbance Variables (DVs) and the Controlled Variables (CVs) have been determined. The setpoint of each furnace zone has been included among MVs (u), while furnace production rate has been entered as a DV (d). Among CVs, zone temperature measurements (y) and the temperature of the billet inside the furnace (b) have been included. Exploiting the defined process variables setup, a global process model has been obtained exploiting different modeling procedures. A step test phase has been executed on the real plant in order to capture the most significant dynamics of the furnace process variables (e.g. zone temperatures measurements) while the first principle approach has been exploited for modeling billets' temperature profile inside the furnace [7]. The designed billets' temperature model acts as a virtual sensor for billets' temperature estimation; in fact no measurement of the billets' temperature inside the furnace is available. Furthermore, the model after proper linearization has been included in the APC formulation.

The main features of the developed APC system are:

- Exploitation of temperature information: the APC system exploits the temperature information given by the optical pyrometers located at the furnace inlet and by the pyrometer located at rolling mill stands;
- Adaptive model: the billets' temperature model is online adapted exploiting feedback measurements by the pyrometer located at rolling mill stands;
- Model Predictive Control (MPC) approach: the APC system, exploiting the feedback information about furnace production rate (future deviations are not a priori known), computes billets' temperature predictions at rolling mill pyrometer;
- Downtime management: the APC system optimizes all furnace conditions (regular conduction and downtime).

Figure 7 shows the scheme of the proposed APC system, based on an MPC strategy [8, 9]. A two-layer MPC strategy is based on the solution of two cascaded optimization problems, solved by a Dynamic Optimizer module (Fig. 7, *DO*) and a Targets Optimizing and Constraints Softening module (Fig. 7, *TOCS*) [10]. *DO* module, i.e. the lower layer of the proposed MPC structure, has been based on the minimization of the quadratic cost function (2), subject to the linear constraints (3).

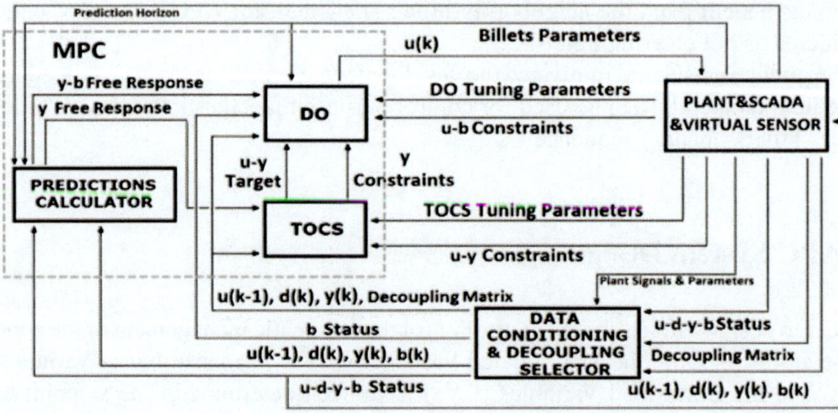

Fig. 7 Scheme of the proposed APC system

$$V_{DO}(k) = \sum_{i=0}^{H_p-1} \|\hat{u}(k+i|k) - u_t(k+i|k)\|^2_{\mathcal{S}(i)} + \sum_{i=1}^{H_p} \|\hat{y}(k+i|k) - y_t(k+i|k)\|^2_{\mathcal{Q}(i)} + \tag{2}$$

$$\sum_{i=1}^{H_u} \|\Delta\hat{u}(k+M_i|k)\|^2_{\mathcal{R}(i)} + \|\varepsilon_y(k)\|^2_{\rho_y} + \sum_{j=1}^{m_b} \left\|\hat{b}_j\left(k+e_j|k\right) - lb_{b\_DO\,j}\right\|^2_{T_j} + \|\varepsilon_b(k)\|^2_{\rho_b}$$

subject to

$$\begin{aligned}
&\text{i.} \quad lb_{du_DO}(i) \le \Delta\hat{u}(k+M_i|k) \le ub_{du_DO}(i),\, i = 1, \ldots H_u \\
&\text{ii.} \quad lb_{u_DO}(i) \le \hat{u}(k+M_i|k) \le ub_{u_DO}(i),\, i - 1, \ldots H_u \\
&\text{iii.} \quad lb_{y_DO}(i) - \gamma_{lby_{DO}}(i) \cdot \varepsilon_y(k) \le \hat{y}(k+i|k) \le ub_{y_{DO}}(i) \\
&\qquad + \gamma_{uby_{DO}}(i) \cdot \varepsilon_y(k),\, i = 1, \ldots H_p \\
&\text{iv.} \quad lb_{b_DOj} - \gamma_{lbb_DOj} \cdot \varepsilon_{bj}(k) \le \hat{b}\left(k+e_j|k\right) \le ub_{b_DOj} \\
&\qquad + \gamma_{ubb_DOj} \cdot \varepsilon_{bj}(k),\, j = 1, \ldots m_p \\
&\text{v.} \quad \varepsilon_y(k) \ge 0;\, \varepsilon_b(k) \ge 0
\end{aligned} \tag{3}$$

DO module computes the control input to be applied to the plant at each control instant (Fig. 7, *u(k)*). The virtual sensor shown in Fig. 7 provides the billets' temperature estimation at each control instant and their adaptive model.

DO module computes the control input to be applied to the plant at each control instant (Fig. 7, *u(k)*). The virtual sensor shown in Fig. 7 provides the billets' temperature estimation at each control instant and their adaptive model.

4 Field Results

The proposed APC system has been installed on the real plant in 2019. Figures 8, 9, 10 and 11 show a performance example (about 14 h). The furnace production rate (Fig. 8) is not constant (range 0–200 t/h) and its deviations are not a priori known by the APC system. The billets' input temperature (Fig. 9), measured by the two optical pyrometers located near the furnace entrance, varies in the range 30 °C–910 °C and its deviations are not a priori known by the APC system. Despite the significant variations of the furnace production rate and of the billets' furnace input temperature, the APC system guarantees a satisfactory reheating of the billets that transit the furnace (Fig. 10, red line represents the lower constraint (990 °C)), through

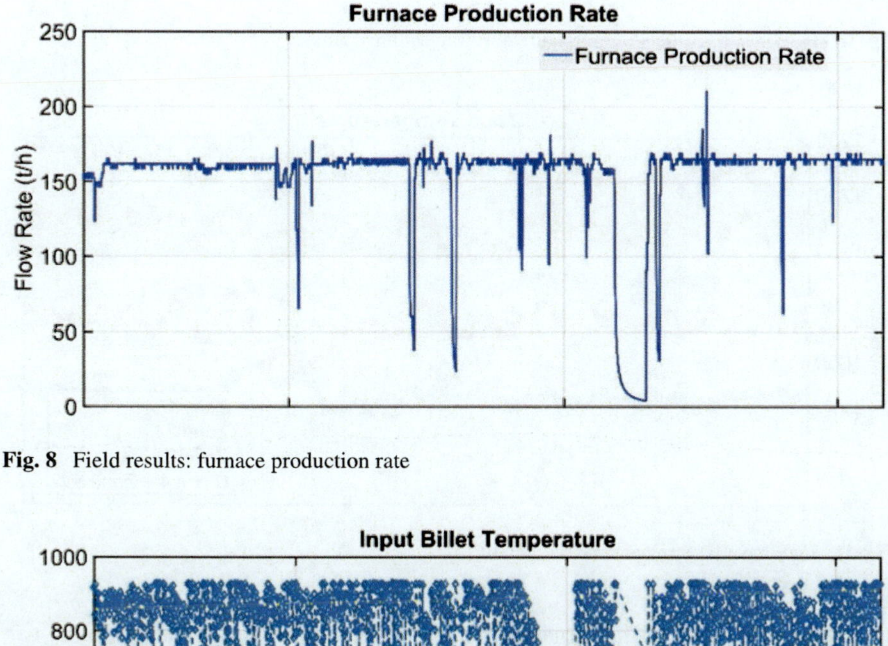

Fig. 8 Field results: furnace production rate

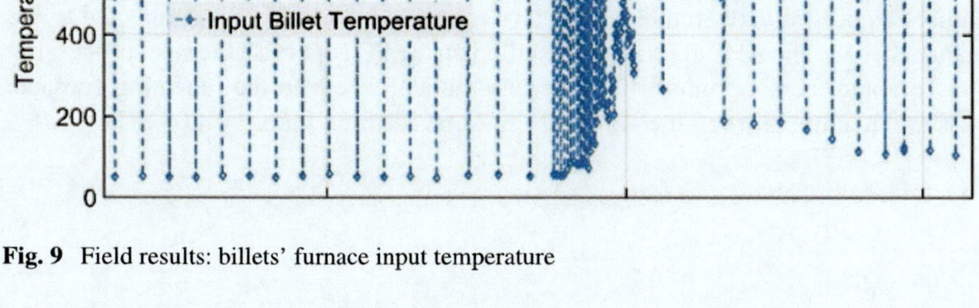

Fig. 9 Field results: billets' furnace input temperature

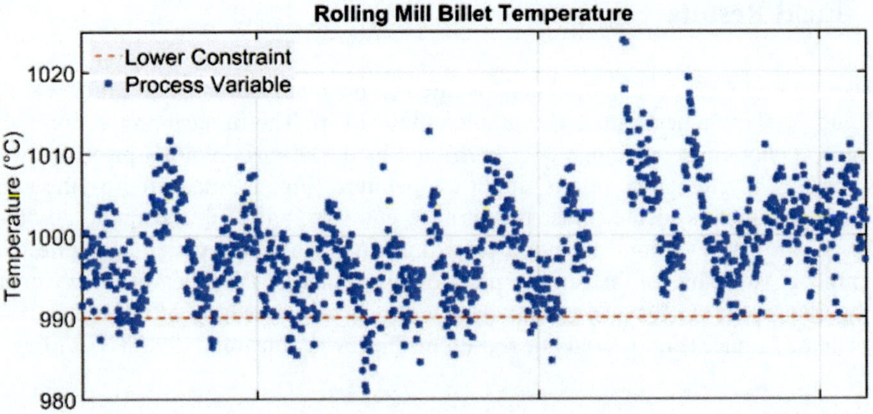

Fig. 10 Field results: billets' temperature at rolling mill pyrometer

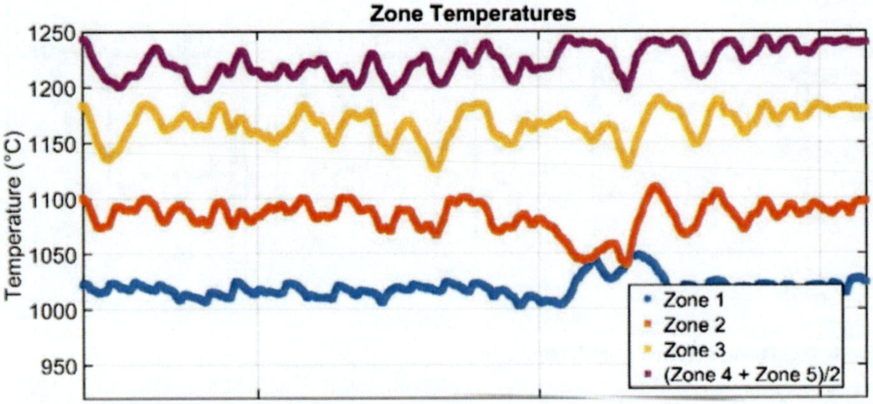

Fig. 11 Field results: furnace zone temperatures

a dynamic regulation of the furnace zone temperatures (Fig. 11). After about a year from the first start-up, the monthly APC system service factor was never less than 85%; furthermore, the automatic management of the zone temperatures has been a remarkable support for the plant operators. The Advanced Process Control system hardware and software architecture is characterized by a PC that is connected to the plant network. The APC algorithms installed on the PC supervise the steel processing and compute the set-points which are automatically sent to the reheating furnace. This architecture allowed the APC system to be certified Industry 4.0 [11].

5 Conclusions

In this work, data analysis and control theory techniques have been applied to design and implement an Advanced Process Control (APC) system on a pusher-type billets' reheating furnace located in an Italian steel plant. Power absorption of the first six rolling stands (roughing phase) downstream of the reheating furnace with respect to process variables and inlet and outlet billet temperature measurements has been analyzed. The application of multivariate regression techniques has been proven satisfactory to predict the energy absorption of the rolling stands. As a first result, from the regressor analysis, a prefiltering of the inlet billet temperature has been considered in the development of the Model Predictive Control system.

The APC system replaced previous totally manual conduction of the furnace zone temperatures and provided a very significant support for plant operators and technicians. A remarkable service factor and good results from fuel specific consumption point of view confirm the validity of the proposed approach.

Industry 4.0 certification has been obtained thanks to the APC architecture and the technical solution has been awarded with an Italian patent.

Currently, in operating the furnace, what operators must take into consideration is not exceeding the pre-established absorption thresholds that would lead to the breaking of the cages. To prevent cages breaking, the operators set suitable constraints values of outlet furnace billet temperature. It turns out, in some conditions, that these set-points are kept excessively high to avoid the possibility of breakages. As the multivariate regression analysis has shown, the outlet rolling mill temperature is not explicative by itself of the different absorption values at the rolling stands. From the above, it emerges that, as a future work, the multivariate regression analysis could be exploited to develop an automated procedure for setting the temperature values supplied to the controller.

References

1. Trinks W, Mawhinney MH, Shannon RA, Reed RJ, Garvey JR (2004) Industrial furnaces. Wiley, New York
2. van Niekerk SGJ, Breytenbach WJJ, Marais JH (2018) In: Proceedings of the international conference on the industrial and commercial use of energy (Cape Town) (IEEE), pp 1–6
3. Tang G, Wu B, Bai D, Wang Y, Bodnar R, Zhou CQ (2017) Int J Heat Mass Transf 113:1142–1151
4. Santos HSO, Almeida PEM, Cardoso RTN (2017) Model Simul Eng
5. Wu M, Cao W, Chen X, She J (2020) Intelligent optimization and control of complex metallurgical processes. Springer, Singapore
6. Steinboeck A (2011) Model-based control and optimization of a continuous slab reheating furnace. Shaker Verlag GmbH, Aachen
7. Astolfi G, Barboni L, Cocchioni F, Pepe C (2016) Metodo per il controllo di forni di riscaldo (Italian Patent n. 0001424136 awarded by Ufficio Italiano Brevetti e Marchi
8. Handbook of model predictive control (2019) Raković SV, Levine WS (eds.), (Birkhäuser Basel)

9. Zanoli SM, Barboni L, Cocchioni F, Pepe C (2018) Advanced process control aimed at energy efficiency improvement in process industries. In: IEEE international conference on industrial technology (ICIT), pp 57–62
10. Pepe C, Zanoli SM (2016) A two-layer model predictive control system with adaptation to variables status values. In: 17th international carpathian control conference (ICCC), pp 573–578
11. Bonci A, Pirani M, Longhi S (2016) A database-centric approach for the modeling, simulation and control of cyberphysical systems in the factory of the future. In: Proceedings of the 8th IFAC conference on manufacturing modelling, management & control, MIM 2016, Troyes, France, June 28–30, pp 249–254

Designing an Active Anti-Roll Bar Control System for a Semi-Trailer

Duc Tien Bui, Olivier Sename, and Van Tan Vu

Abstract Nowadays, tractor semi-trailer heavy vehicles are the most popular transport vehicle in the world because of their benefits. However, the traffic accidents related to these vehicles usually are serious consequences. The main reason causing these accidents is the loss stability of tractor semi-trailer heavy vehicles, particularly the rollover phenomenon of them. In order to improve the roll stability most modern, these vehicles have equipped a passive anti-roll bar system. However, the passive anti-roll bar does not meet the required stability when the vehicle is in emergency. This paper investigates the influence of the suspension roll damping and roll stiffness parameters on the stability of the tractor semi-trailer and designs an active anti-roll bar control system for semi-trailer, by using Linear Quadratic Regulator (LQR) method. The obtained results show that these suspension parameters do not meet well for the vehicle roll stability when the tractor semi-trailer is in a cornering maneuver. In addition, the simulation results compare the performance between the passive and the LQR active anti-roll bar systems for semi-trailer which show that by using the active system the reduction for the roll angle of the sprung mass is about 43%, 50% for the normalized load transfer, and 50% for the roll angle of the unsprung mass.

Keywords Tractor semi-trailer · Active anti-roll bar · LQR control method · Roll stability · Rollover

D. T. Bui (✉)
Department of Automotive Engineering, Faculty of Mechanical Engineering, Thuyloi University, 175 Tay Son, Dong Da, Hanoi, Vietnam
e-mail: buiductien.ckoto@tlu.edu.vn

O. Sename
GIPSA-Lab, Institute of Engineering Univ. Grenoble Alpes, Univ. Grenoble Alpes, CNRS, Grenoble INP, 38000 Grenoble, France

V. T. Vu
Department of Automotive Mechanical Engineering, Faculty of Mechanical Engineering, University of Transport and Communications, Hanoi, Vietnam

G. Conte and O. Sename (eds.), *Proceedings of 10th International Conference on Mechatronics and Control Engineering*, Lecture Notes in Mechanical Engineering,
https://doi.org/10.1007/978-981-19-1540-6_16

1 Introduction

Today, tractor semi-trailer heavy vehicles play an essential role in the transportation sector because of their merits. Compared to conventional trucks, tractor semi-trailer vehicles provide such benefits as reduction in fuel consumption per ton per kilometer leading to less emission, reduce transport costs thanks to increase in the transport volume and payload of one delivery time, and road protection due to reduced axle loads.

However, the accidents related to tractor semi-trailer usually are fatal and they often are caused by rollover accidents. There are three main factors leading to rollover accidents which are side wind gusts, abrupt steering, and braking maneuvers by the driver. In these cases, roll stability occurs when the tire–road contact force of the wheels on one side of the vehicle equals zero. Hence, most of the tractor semi-trailers are equipped with passive anti-roll bar at all axles in order to cope with the instability of tractor semi-trailer. However, it also has some drawbacks: the passive anti-roll bar system will transfer vertical forces of one side of the suspension to other side and leading moment to against the lateral force when driver control is in tractor semi-trailer cornering situation.

An active anti-roll bar system is applied to deal with these drawbacks. The active anti-roll bar includes a pair of hydraulic actuators [1], it generates a stabilizing moment to balance the overturning moment. Therefore, the lateral load response can reduce significantly.

The research on controlling the active roll control system for the tractor semi-trailer was conducted by a research team at the University of Cambridge in the United Kingdom. In [2], A.J.P. Miège and D. Cebon used PID control method to apply on an actual tractor semi-trailer, and the authors had controlled the active ARB system to improve the vehicle roll stability. A semi-active anti-roll bar system is proposed by combining active and passive systems in [3]. In [4, 5], the authors designed a state feedback roll control system for a tractor semi-trailer using a flexible frame, which allowed a more accurate assessment of the benefits of this system. The mechanical properties of tire effect on instability of tractor-trailer at moderate levels of lateral acceleration are indicated in [6].

The previous studies proposed an active anti-roll bar system for both tractor and semi-trailer, while Sampson's results showed that the rear axle of semi-trailer is loss stability first [1]. Therefore, the objective control for semi-trailer is easier and exchangeable with other tractor as well as improve significantly the stability of tractor semi-trailer vehicles. Therefore, this is a new approach and more practical solution when compared with the previous studies.

The contributions of this paper are as follows:

- A yaw–roll model of tractor semi-trailer is used for surveying the influence of structural parameters on the stability of the tractor semi-trailer, by considering the vertical and lateral displacement of the vehicle.
- An optimal LQR controller is designed at the semi-trailer axle in order to improve the roll stability of tractor semi-trailer.

- The simulation results in time domain and frequency domain show that the LQR active anti-roll bar system improves significantly the roll stability of tractor semi-trailer when compared with the passive anti-roll bar system.

2 Vehicle Modeling

2.1 Modeling of a Tractor Semi-Trailer

In this study, a yaw–roll model of tractor-semi-trailer in [7] is used in order to analyze characteristic dynamics with some assumptions: the tractor-semi-trailer is assumed to be perfectly rigid; pitching and bouncing motions have only a small effect on the roll and handling behavior of the vehicle and so can be neglected in formulating a model to investigate roll and handling performance. The effects of aerodynamic inputs (wind disturbances) and road inputs (cross-gradients, dips and bumps) are also neglected.

The vehicle body has the roll axis with distance r upward from the ground. The dynamic equations of the vehicle are calculated by equating the change of momentum (or, in the rotational case, moment of momentum) with the sum of external forces (or moments) acting on the system. The motion is described using a coordinate system (x', y', z') fixed in the vehicle, as shown in Fig. 1. The roll axis is replaced by an

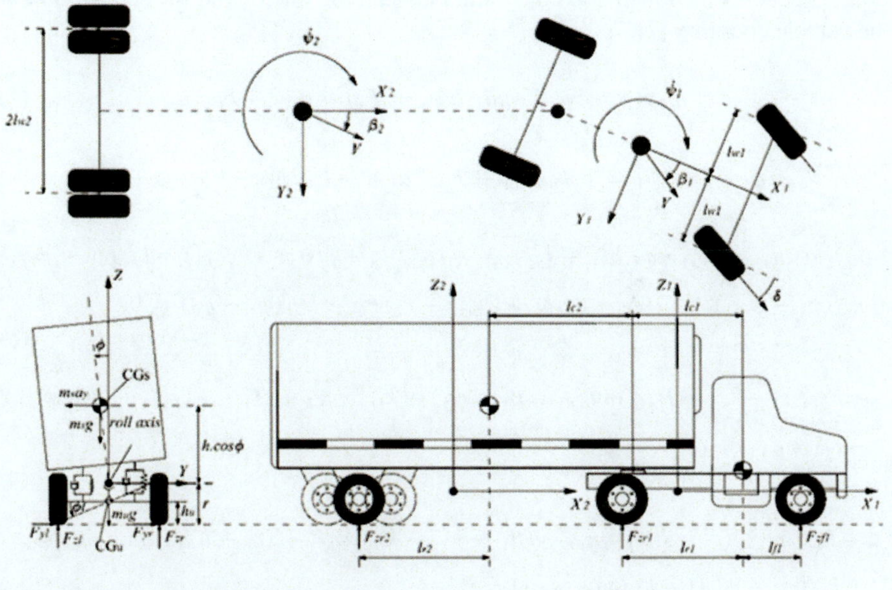

Fig. 1 Yaw–roll model of tractor-semi-trailer

x'-axis parallel to the ground and the z'-axis passes downward through the center of mass of the tractor-semi-trailer.

Figure 1 shows the tractor-semi-trailer model which includes two parts: tractor and semi-trailer, in which is the sprung mass, is the unsprung mass at the front axle, and the unsprung mass at the rear axle, is the height of center of sprung mass, is the height of center of unsprung mass, is the roll moment of inertia of sprung mass, is the yaw moment of inertia of sprung mass, is the yaw roll moment of inertia of sprung mass, is the longitudinal distance to articulation point, measured forward from center of sprung mass, is the longitudinal distance to articulation point, measured forward from center of total mass of tractor or semi-trailer. The subscripts f and r express the front and rear axles of the tractor or trailer, respectively. The subscripts i (1 and 2) express the tractor and semi-trailer, respectively. Is lateral force at the coupling point. The symbols and parameters of this model are shown in [7].

2.2 *Dynamic Equations of a Tractor-Semi-Trailer*

Equations (1)–(10) show the dynamic of the yaw–roll model of tractor-semi-trailer, in which the yaw moment, the roll moment of the sprung mass, and the roll moment of the front and rear unsprung masses of tractor are expressed in Eqs. (1)–(5). Equations (6)–(8) describe the yaw moment of the sprung mass, the roll moment of the sprung mass, and the roll moment of the rear unsprung mass of semi-trailer. Equation (9) denotes the kinematic constraint between the tractor and the semi-trailer at the vehicle coupling [7]:

$$m_1 v(\dot{\beta}_1 + \dot{\psi}_1) - m_{s1} h \dot{\phi}_1 = F_{yf} + F_{yr} - F_C \tag{1}$$

$$I_{xz1}.\ddot{\phi}_1 - I_{zz1}.\ddot{\psi}_1 = F_{yr1}.dl_{r1} - F_{yf1}.dl_{f1} - b'_{r1}.F_C - k_{\psi 1}(\psi_1 - \psi_2) \tag{2}$$

$$\begin{aligned} &-m_{s1}.a_y.h_{s1} - m_{s1} g.\phi_1.h_{s1} + I_{xx1}.\ddot{\phi}_1 - I_{xz1}.\ddot{\psi}_1 + k_{f1}(\phi_1 - \phi_{tf1}) + l_{f1}(\dot{\phi}_1 - \dot{\phi}_{tf1}) \\ &+k_{r1}(\phi_1 - \phi_{tr1}) + l_{r1}(\dot{\phi}_1 - \dot{\phi}_{tr1}) + (h_{ar1} - r_1)F_C - k_{\phi 1}(\phi_1 - \phi_2) = 0 \end{aligned} \tag{3}$$

$$\begin{aligned} &-F_{yf1}.r_1 + (r_1 - h_{uf1}).m_{uf1}.v(\dot{\beta}_1 + \dot{\psi}_1) + k_{tf1}.\phi_{tf1} + (r_1 - h_{uf1}).m_{uf1}.g.\phi_{tf1} \\ &-k_{f1}(\phi_1 - \phi_{tf1}) - l_{f1}(\dot{\phi}_1 - \dot{\phi}_{tf1}) = 0 \end{aligned} \tag{4}$$

$$\begin{aligned} &-F_{yr1}.r_1 + (r_1 - h_{ur1}).m_{ur1}.v(\dot{\beta}_1 + \dot{\psi}_1) + k_{tr1}.\phi_{tf1} + (r_1 - h_{uf1}).m_{ur1}.g.\phi_{tr1} \\ &-k_{r1}(\phi_1 - \phi_{tr1}) - l_{r1}(\dot{\phi}_1 - \dot{\phi}_{tr1}) = 0 \end{aligned} \tag{5}$$

$$I_{xz2}.\ddot{\phi}_2 - I_{zz2}.\ddot{\psi}_2 = F_{yr2}.dl_{r2} + b'_{f2}.F_C - k_{\psi 1}(\psi_1 - \psi_2) \tag{6}$$

$$\begin{aligned} & I_{xx2}.\ddot{\phi}_2 - I_{xz2}.\ddot{\psi}_2 - m_{s2}.a_y.h_{s2} - m_{s2}g.\phi_2.h_{s2} + k_{r2}(\phi_2 - \phi_{tr2}) \\ & + l_{r2}(\dot{\phi}_2 - \dot{\phi}_{tr2}) + u_{r2} - (h_{af2} - r_2)F_C + k_{\phi 1}(\phi_1 - \phi_2) = 0 \end{aligned} \tag{7}$$

$$\begin{aligned} & -F_{yr2}.r_2 - (h_{ur2} - r_2).m_{ur2}.v(\dot{\beta}_2 + \dot{\psi}_2) - (h_{ur2} - r_2).m_{ur2}.g.\phi_{tr2} \\ & -k_{r2}(\phi_2 - \phi_{tr2}) - l_{r1}(\dot{\phi}_2 - \dot{\phi}_{tr2}) + k_{tr2}.\phi_{tr2} = 0 \end{aligned} \tag{8}$$

$$\beta_1 - \beta_2 - \frac{(r_1 - h_{ar1})}{v}\dot{\phi}_1 + \frac{(r_2 - h_{af2})}{v}\dot{\phi}_2 + \frac{b'_{r1}}{v}\dot{\psi}_1 - \frac{b'_{f2}}{v}\dot{\psi}_2 + \psi_1 - \psi_2 = 0 \tag{9}$$

The lateral tire forces in the direction of velocity at the wheel ground connection points which are modeled by using linear stiffness coefficients as follows:

$$\begin{aligned} F_{yfi} &= \mu C_f \alpha_f \\ F_{yri} &= \mu C_r \alpha_r \end{aligned} \tag{10}$$

With the tire side slip angles:

$$\alpha_{f1} = -\beta_1 + \delta_f - \frac{l_{f1}\dot{\psi}_1}{v} \qquad \alpha_{r1} = -\beta_1 + \frac{l_{r1}\dot{\psi}_1}{v} \qquad \alpha_{r2} = -\beta_2 + \frac{l_{r2}\dot{\psi}_2}{v} \tag{11}$$

The tractor semi-trailer is considered rollover when the wheel lifts off, which means the normalized load transfer reaches ± 1 [7].

3 Influence of Structural Parameters on Tractor-Semi-Trailer Stability

In this part, we use a cornering situation [8] as the common disturbance for studying the roll stability when the vehicle velocity is considered at 60 km/h. Figure 2 shows the time response of the roll angle of sprung mass when the characteristic of suspension roll damping and suspension roll stiffness changes from (0.5 ÷ 1.5) standard value in, especially ($4.10^5 \div 12.10^5$) Nm for suspension roll stiffness and ($5.10^4 \div 15.10^5$) Nm/rad for suspension roll damping.

As we see, the less value of suspensison roll stiffness and suspension roll damping are the roll angle of the sprung mass of semi-trailer and reaches the peak at 15, 28° when the value of suspensison roll stiffness and suspension roll damping are the lowest. When these figures increase, the roll angle of the sprung mass of semi-trailer is decreased, respectively, and it gets the lowest value at 6, 03°. We also see that this roll angle of the sprung mass depends mainly on the suspension roll stiffness, and the suspension roll damping affects it insignificantly.

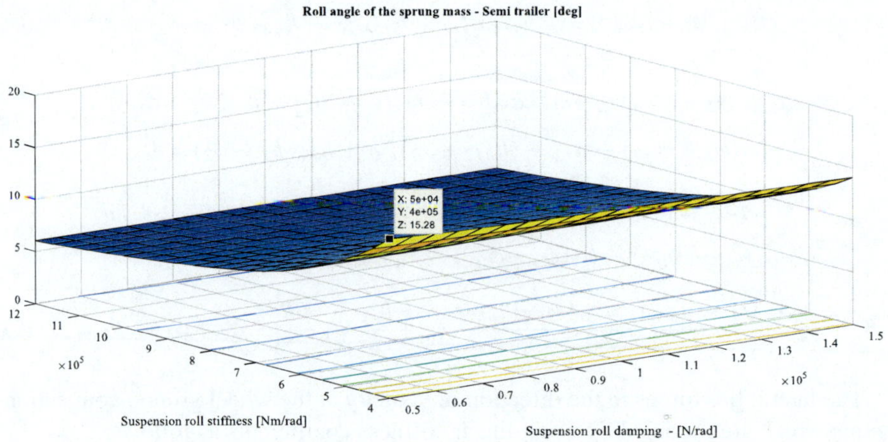

Fig. 2 Time responses of the roll angle of the sprung mass

Figure 3 shows the time response of the normalized load transfer at rear axles of semi-trailer. We can see that when the value of suspensison roll stiffness and suspension roll damping are low, the normalized load transfer is low and reaches the lowest point at 0.93. This value rises to 1.21 when the value of suspensison roll stiffness and suspension roll damping increase. However, the normalized load transfer also rounds to 1. It is clear that the risk of rollover of this vehicle is usually high when the velocity is at 60 km/h although the value of suspensison roll stiffness and suspension roll damping vary. Therefore, designing an active anti-roll bar system controller for tractor semi-trailer in order to improve the stability of this vehicle is very neccessary.

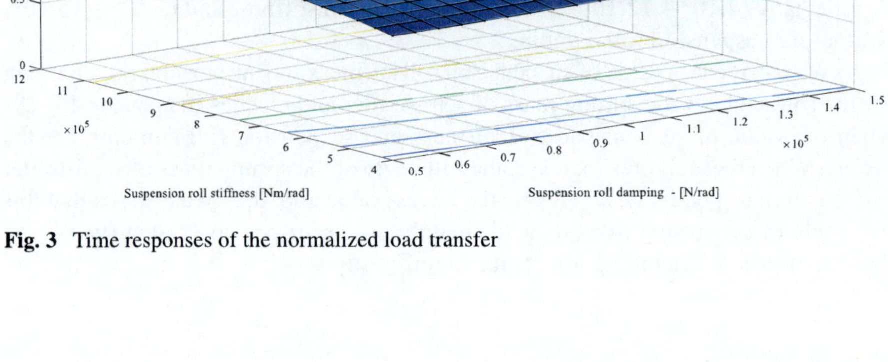

Fig. 3 Time responses of the normalized load transfer

4 LQR Controller Design for Active Anti-Roll Bar System of Semi-Trailer

The motion differential equations from (1) to (11) can be rewritten in the LTI state-space representation as follows:

$$\begin{cases} \dot{x} = Ax + B_1 w + B_2 u \\ z = Cx + D_1 w + D_2 u \end{cases} \tag{12}$$

where

the state vector: $x = [\beta_1 \ \ \dot{\psi}_1 \ \ \phi_1 \ \ \dot{\phi}_1 \ \phi_{tf1} \ \phi_{tr1} \ \phi_2 \ \ \dot{\phi}_2 \ \ \beta_2 \ \ \dot{\psi}_2 \ \ \phi_{tr2}]^T$,

the exogenous disturbance: $w = [\delta\,]^T$,

the control input: $u = [\,u_{r2}]^T$, and

the output vector: $z = x$.

4.1 The Control Objectives for Vehicle Roll Stability

The objective of the active anti-roll bar control system is to maximize the vehicle roll stability. In this paper, we would like to minimize the normalized load transfer of rear axle of semi-trailer to evaluate the vehicle roll stability of tractor semi-trailer vehicles:

$$|R| = \left| \frac{\Delta F_z}{F_z} \right| < 1 \tag{13}$$

where ΔF_z is lateral load transfer and F_z total axle load of semi-trailer. If R takes the value ± 1, the rollover occurs because the inner wheel in the bend lifts off and denoted in [8].

4.2 LQR Active Anti-Roll Bar Controller

The LTI model is noted by Eq. (12). For the controller design, it is assumed that all the states are available from measurements or can be estimated. Then, we consider the state feedback control law:

$$u = -Kx \tag{14}$$

where K is the state feedback gain matrix. The optimization procedure includes determining the control input u which minimizes some performance index J. This index includes the performance characteristic requirement as well as the controller input limitations, usually expressed by $J = \int_0^\infty (x^T Qx + u^T Ru)dt$, where Q and R are the positive definite weighing matrices. The optimal closed-loop system is obtained as follows [9]:

$$\dot{x} = (A - B_2K)x + B_1W \tag{15}$$

The performance index J is chosen as follows:

$$J = \int_0^\infty (\rho_1\phi_2^2 + \rho_2\,\dot{\phi}_2^2 + \rho_3\beta_2^2 + \rho_4\,\dot{\psi}_2^2 + \rho_5\phi_{tr2}^2)dt \tag{16}$$

Remark The selection of J completely depends on the goal of the control design. $\rho_1 = 10^{13}$; $\rho_2 = 10$; $\rho_3 = 10$; $\rho_4 = 10$; $\rho_5 = 10$ are the weighting parameters of J.

4.3 The Simulation Results in the Time Domain

In this paper, we use a cornering situation [8] as the common disturbance for studying the roll stability. Figure 4 shows the time response of the steering angle, the roll angle of sprung mass, the roll angle of unsprung mass, and the normalized load transfer at rear axles of semi-trailer when the vehicle velocity is considered at 60 km/h. We can see that the maximum value of these variables in LQR controller is always smaller than the passive anti-roll bar system. The reduction is about 43% for the roll angle of the sprung mass, 50% for the normalized load transfer, and 50% for the roll angle of the unsprung mass.

4.4 The Simulation Results in the Frequency Domain

We simulate tractor-semi-trailer model in the frequency range from $10^{-2} \div 10^2$ rad/s with the control input is the steering angle. Figure 5 illustrates the transfer from steering angle to the roll angle of the sprung mass of semi-trailer. We can see this value reduces to 4 dB in the frequency range of $10^{-2} \div 4$ rad/s, which is respected for the driver bandwidth [8, 9]. We can conclude that by using LQR active anti-roll bar control system at the rear axle of semi-trailer, the roll stability of the tractor semi-trailer vehicles is significantly improved when compared with a passive system.

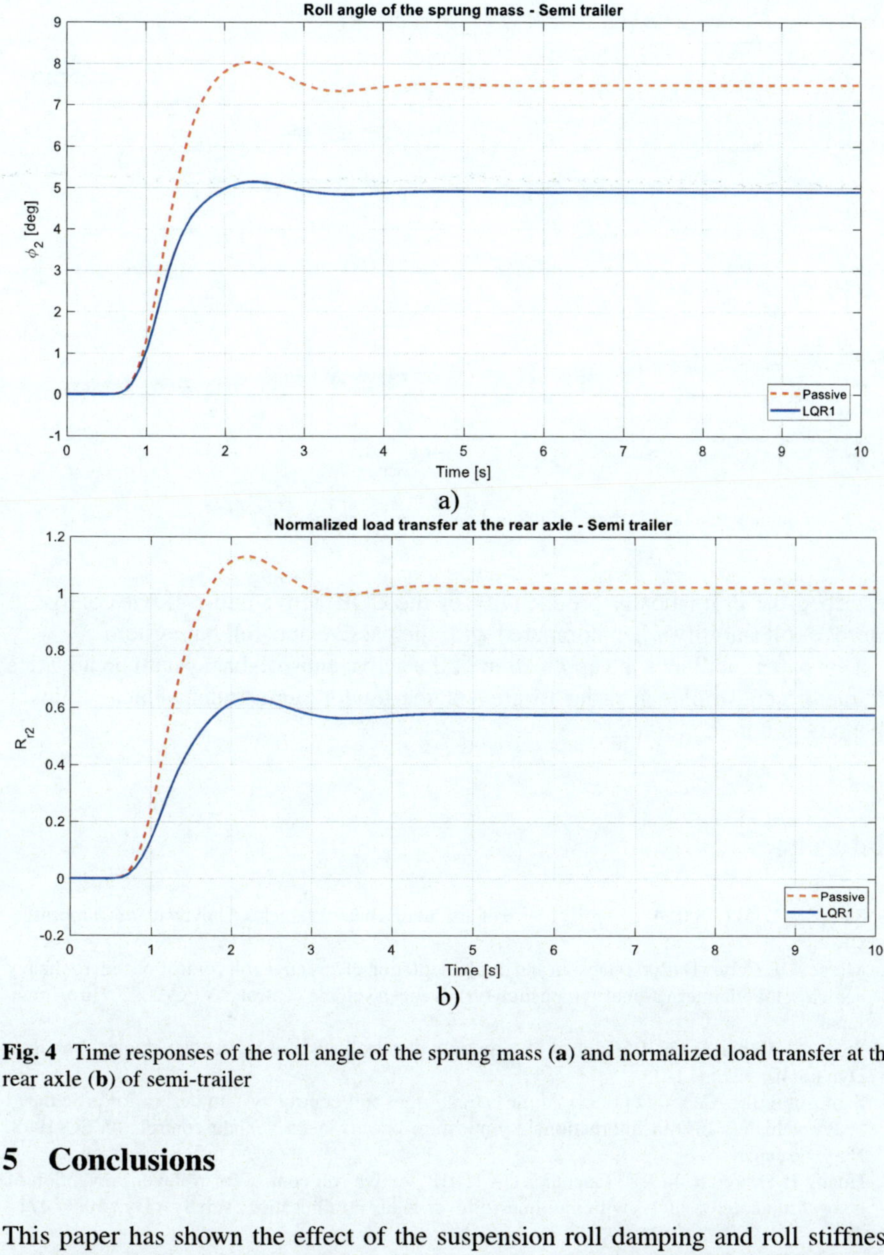

Fig. 4 Time responses of the roll angle of the sprung mass (**a**) and normalized load transfer at the rear axle (**b**) of semi-trailer

5 Conclusions

This paper has shown the effect of the suspension roll damping and roll stiffness parameters on the stability of the tractor semi-trailer and illustrated merits of the anti-roll bar system on tractor semi-trailer, particularly on semi-trailer during cornering situations. The LQR controller is applied on the rear axle of the semi-trailer, in which the lateral load transfer is taken into consideration. The obtained results have

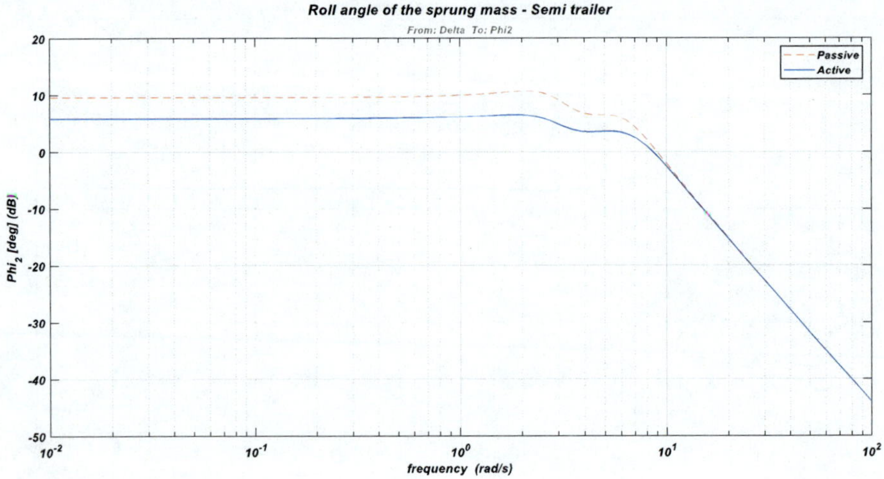

Fig. 5 Frequency responses when vehicle is in a cornering maneuver

described the dramatically productivity of the LQR active anti-roll bar control to improve roll stability when compared with the passive anti-roll bar system.

This paper facilitates the application of the active anti-roll bar system on all axles of tractor semi-trailer in order to prevent the tractor semi-trailer vehicle rollover phenomenon in the future.

References

1. Sampson DJM (2000) Active roll control of articulated heavy vehicles. University of Cambridge, UK
2. Miège AJP, Cebon D (2002) Design and implementation of an active roll control system for heavy vehicles. In: 5th international symposium on advanced vehicle control, AVEC-2002, Hiroshima, Japan
3. Stone EJ, Cebon D (2010) Control of semi-active antiroll systems on heavy vehicles. Veh Syst Dyn 48(10):1215–1243
4. Sampson DJM, Cebon D (1998) An investigation of roll control system design for articulated heavy vehicles. In: 4th international symposium on advanced vehicle control, AVEC-1998, Nagoya, Japan
5. Huang H-H, Yedavalli RK, Guenther DA (2012) Active roll control for rollover prevention of heavy articulated vehicles with multiple-rollover-index minimisation. Veh Syst Dyn 50(3):471–493
6. Segel L, Ervin RD (1981) The influence of tire factors on the stability of trucks and truck-trailers. Veh Syst Dyn 10(1):39–59
7. Tan VV, Sename O, Gáspár P (2021) Improving roll stability of tractor semi-trailer vehicles by using H∞ active anti-roll bar control system. In: Proceedings of the institution of mechanical engineers. Part D: Journal of Automobile Engineering, Published online first May 08
8. Tan VV, Hung ND (2020) Using an LQR active anti-roll bar system to improve road safety of tractor semi-trailers. Sci Technol Dev J 23(3):593–601

9. Gaspar P, Szabo Z, Bokor J (2004) Prediction based combined control to prevent the rollover of heavy. Limassol, Cyprus. In: Proceedings of the 13th mediterranean conference on control and automation 10(2):148–162

Author Index

G. Conte and O. Sename (eds.), *Proceedings of 10th International Conference on Mechatronics and Control Engineering*, Lecture Notes in Mechanical Engineering, https://doi.org/10.1007/978-981-19-1540-6